Asphalt Core Embankment Dams

Dams are built for several purposes, e.g., water supply, irrigation, hydropower production, or flood control. The main categories are concrete dams and embankment dams, but inside each of these two main general categories, there are several different types.

Asphalt Core Embankment Dams focusses on the design and construction of these types of dams. Asphalt core embankment dams are gaining in popularity, as their recorded field performance is excellent, they are economically very competitive, and the construction period may be shortened compared to other types of embankment dams. This book presents several case studies of dams that have been successfully built and been in service for many years, and:

- introduces the concept of using asphalt concrete core rather than earth core or upstream facing;
- points out there is no concern about internal erosion when using hydraulic asphalt concrete in the core;
- discusses material selection and mix design for hydraulic asphalt concrete to achieve flexibility and ductility:
- discusses suitability testing and asphalt concrete mix improvements and optimization by using additives;
- presents the use of triaxial tests, tension tests, and bending tests to study cracking resistance and potential self-healing;
- discusses the use of modern construction equipment and field quality control;
- shows that asphalt core dams may reduce the construction period compared to other options; and
- treats earthquake resistance of this type of dams.

The authors have built on decades of experience with the design and construction, construction control, and research and development for asphalt core embankment dams. *Asphalt Core Embankment Dams* is aimed at engineering students, practising engineers, civil engineering contractors, equipment suppliers, and asphalt producers.

Asphalt Core Embankment Dams

Kaare Høeg and Weibiao Wang

CRC Press
Taylor & Francis Group
Boca Raton London New York Leiden

CRC Press is an imprint of the
Taylor & Francis Group, an **informa** business
A BALKEMA BOOK

Front cover image: Storglomvatn Dam by the Arctic Circle, Norway. Courtesy of Statkraft AS, Norway

First published 2025
by CRC Press/Balkema
4 Park Square, Milton Park, Abingdon, Oxon, OX14 4RN

and by CRC Press/Balkema
2385 NW Executive Center Drive, Suite 320, Boca Raton FL 33431

CRC Press/Balkema is an imprint of the Taylor & Francis Group, an informa business

British Library Cataloguing-in-Publication Data
A catalogue record for this book is available from the British Library

ISBN: 978-1-138-03763-2 (hbk)
ISBN: 978-1-041-02728-7 (pbk)
ISBN: 978-1-315-17778-6 (ebk)

DOI: 10.1201/9781315177786

Typeset in Times New Roman
by codeMantra

Contents

Acknowledgements

Norwegian Geotechnical Institute (NGI), Norway; Xi'an Huize Construction Engineering Co. Ltd (XI'ANHUIZE), China; and Xi'an University of Technology (XAUT), China, are gratefully acknowledged for supporting the research and development performed by the authors presented in this book.

Special thanks go to Lang Pu for her expert help in drawing most of the figures presented in this book.

Author biographies

Kaare Høeg is expert advisor at the Norwegian Geotechnical Institute (NGI) and a professor emeritus in the Department of Geosciences at the University of Oslo. He earned his doctorate (ScD) from the Massachusetts Institute of Technology (MIT), USA, and was a professor at Stanford University, USA, from 1968 until 1974 when he was appointed the Managing Director of NGI, a position he held until 1991. He was elected as a member of the Norwegian Academy of Technical Sciences and the Academy of Letters and Arts and as a foreign member of the US National Academy of Engineering. He was elected as the president of the International Commission on Large Dams (ICOLD) from 1997 to 2000. He is actively engaged as a design consultant for new dams and as an expert member for the safety evaluation, risk analysis, and strengthening of existing dams. He is the author/co-author of two books on embankment dams and has published 170 technical papers in international journals and conference proceedings.

Weibiao Wang is a supervisor at Xi'an Huize Construction Engineering Co. Ltd and a professor emeritus at Xi'an University of Technology (XAUT), China. He obtained his MSc from XAUT and his doctorate (PhD) from the University of Oslo, Norway. From 1998 to 1999, he was a visiting scholar at the Norwegian Geotechnical Institute (NGI) and stayed at NGI for his PhD studies supervised by Professor Kaare Høeg and later as a postdoctoral researcher from 2007 to 2009. Since 1985, he has been engaged in research and consulting for most of the hydraulic asphalt concrete projects in China. He has co-authored the Chinese NB/T 11015-2022 "Code for design of asphalt concrete facings and cores for embankment dams", the DL/T 5362-2018 "Test code for hydraulic bitumen concrete", and SL 514-2013 "Specifications for construction of hydraulic asphalt concrete". He was one of the four people in the working team for editing and updating the International Commission on Large Dams (ICOLD) Bulletin 179 for asphalt concrete core dams. He is the author/co-author of two books on hydraulic asphalt concrete and has published 60 papers in international journals and conference proceedings.

Chapter 1

Introduction and background

The application of bitumen (asphalt binder) and asphalt concrete in impervious barriers in hydraulic structures may be traced back more than 1,000 years. Systematic development and improvement of hydraulic asphalt concrete (HAC) mixtures has taken place since the 1920s. In the early years, HAC was primarily used as impervious facings for embankment dams and as bottom linings for reservoirs. Asphalt concrete or asphalt mastic membranes have also often been used to seal concrete dams, as linings for canals, and for sea defence works.

The experience gained from asphalt facings, supported by laboratory and field research, led to HAC also being used in impervious core walls in embankment dams. The concept was developed, and the first asphalt core embankment dams (ACEDs) were designed in the 1950s and constructed in the early 1960s. This type of dam has since then proved to be a very competitive design option with respect to safety, speed of construction, and economy.

The basis for the application of asphaltic barriers in dams is the viscoelastic behaviour of HAC. It exhibits flexible and ductile material behaviour, relieves itself of stress and strain concentrations, allows deformations without crack formation, and has self-sealing properties if cracks should develop. An added advantage is the uncomplicated and efficient construction procedure for an ACED and that it can accept inclement weather conditions during construction.

Figure 1.1 shows a photo of the Storglomvatn asphalt core dam in Norway, 128 m high (Høeg et al., 2007). The highest dams of this type built so far are the Quxue Dam (174 m) in Sichuan Province, China (Wang et al., 2017b), and the Moglicé Dam (167 m) in Albania (Tirunas and Aspen, 2023). Most of the existing dams of this type are rockfill dams (asphalt core rockfill dam, ACRD), but the concept and design are suitable for well-compacted embankment dams in general (ACED).

Figure 1.2 shows a typical cross section and zoning of an ACRD.

During the first years of ACED design and construction, German and Austrian dam engineers were the leaders in developing design procedures, specialized asphalt core pavers, and related construction equipment to build the core, and they designed and constructed many of the early ACEDs internationally. Accordingly, the first two International Commission on Large Dams (ICOLD) Bulletins on ACEDs were drafted by the German National Committee of ICOLD (ICOLD, 1982; ICOLD, 1992). Schönian (1999), in an exceptionally well-illustrated book, presented an extensive treatment and documented applications with asphalt concrete used in upstream dam facings and in dam cores.

Since the early 1970s, the construction technology for ACEDs has been applied in Norway, replacing the traditional moraine core. With the help of laboratory investigations, analyses, and field monitoring, Norwegian engineers and contractors have built about 40 ACEDs in Norway

DOI: 10.1201/9781315177786-1

Figure 1.1 Storglomvatn Dam, Norway – close to completion in 1997 (Courtesy of Statkraft).

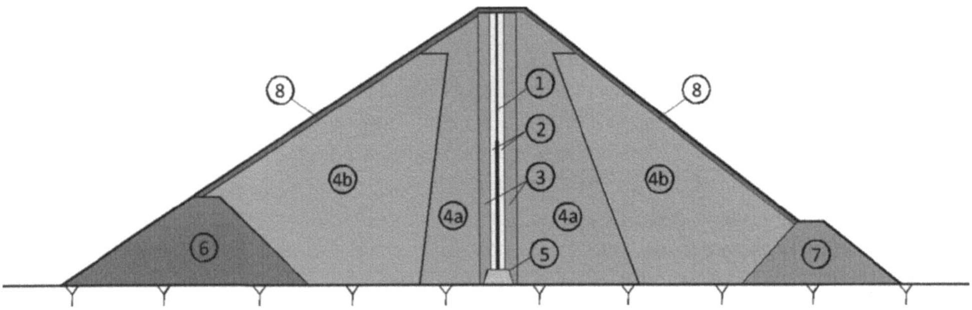

① Asphalt core; ② fine transition zone; ③ coarse transition zone; (4a) well-compacted quarried rockfill; (4b) shoulder of compacted rockfill; ⑤ concrete plinth; ⑥ upstream cofferdam; ⑦ toe drain with stable rocks and boulders; ⑧ riprap protection.

Figure 1.2 Typical design and zone designations for an ACRD on rock foundation.

and abroad. The 128-m high Storglomvatn ACED located by the Arctic Circle in Norway was completed in 1997 and was at that time the highest ACED.

China started building ACEDs in the early 1970s, but the development was slow during the first two decades due to the relatively poor quality of the bitumen available in China at the time and the lack of suitable and effective construction equipment. However, along with the rapid general economic development, China has built more than 100 ACEDs during the last 30 years, and many more are currently under planning, design, and construction. The 125-m high Yele ACED, founded at a geologically complex site in a very seismic region, was completed in 2005. The 174-m high Quxue ACED located in a narrow canyon in Sichuan Province was completed in 2017 (Figure 1.3).

Figure 1.3 The Quxue ACED in China was completed in 2017.

In addition to Europe and China, Japan, Canada, Brazil, Turkey, the Middle East, and African and ASEAN countries have built ACEDs. Creegan and Monismith (1996) emphasized that the properties of asphalt concrete make the material ideally suitable for creating impervious and flexible barriers in dams and recommended the use of ACEDs in North and South America. Canada (Hydro-Québec) has built several ACEDs, and the largest Romaine-2 Main Dam, 112 m high, was completed in 2017. Figures 1.4 and 1.5 show the main dam and the 84-m high saddle dam (Dyke F) for the Romaine-2 Reservoir.

The right abutment of the 84-m high saddle dam for the Romaine-2 Reservoir (Dyke F, Figure 1.5) is very steep, jointed, and fractured. It required special treatment (white area in the photo) to ensure satisfactory core-rock interface contact and to prevent rock blocks from loosening during dam construction. As shown in the photo, the dam crest is widened over a short distance close to the right abutment to allow U-turns of construction vehicles.

The first ACED in the USA was completed in 2021 (Silver Basin Dam, Arizona), and the 110-m high Chimney Hollow Dam, Colorado, is to be completed in 2025 (Figure 1.6).

The performance of the ACEDs built in various countries has been reported to be very satisfactory, and dam experts at the 16th ICOLD Congress (ICOLD, 1992) concluded that the rockfill dam with an asphalt core is an appropriate dam type for the very highest future dams. The first author (Høeg, 1993) summarized some of the experiences from the design and construction of asphalt core rockfill dams up until the early 1990s. The authors have now expanded on that early work by emphasizing design examples and recommendations.

Field monitoring with the accompanying analyses and interpretation of results has been instrumental in advancing the state of the art and has consolidated the profession's understanding of and confidence in ACED behaviour. In China's plans for the very extensive dam-building program in the years ahead, the ACED is playing an important role. That program is expected to further advance the design features and construction procedures for ACEDs.

Figure 1.4 Main dam for the Romaine-2 Reservoir, Canada (Courtesy of Vlad Alicescu, Hydro-Québec).

Figure 1.5 Saddle dam (Dyke F, 84 m high) for the Romaine-2 Reservoir (Courtesy of Jean-Pierre Tournier, Hydro-Québec).

Figure 1.6 Chimney Hollow Dam, Colorado, USA – under construction to be completed in 2025 (Courtesy of WALO International AG-Department-Hydraulic Asphalt Engineering).

In addition to design analyses and recommendations, this book includes data from extensive laboratory and field testing of asphalt concrete for several ACEDs that have been built. The asphalt concrete stress-strain-strength behaviour and parameters that may be used in numerical stress-deformation analyses are discussed as a function of strain rate and temperature, with an emphasis on long-term creep.

The asphalt core presented and discussed in this book is constructed of "dense" grade HAC, compacted in layers 200–300 mm thick. The adjacent gravel transition zones, placed and compacted in the same operation as the core, provide immediate lateral support to the hot core and are essential features of the ACED construction methodology and technology.

Two other methods of asphalt core construction have been used in a few cases, i.e., the "stone-bitumen method", and the use of "flowing" asphalt concrete with a high bitumen content poured and cast-in-place between formworks. In the stone-bitumen method, gravel of suitable quality and gradation is placed between formworks, and then, hot bitumen with fines (filler) is poured over the gravel to fill the voids. This method is no longer recommended because the quality control with that method is not considered satisfactory. The "flowing" asphalt concrete method has been used for a few rockfill dams in Russia where the temperature was extremely low during construction, and the cooling of the asphalt, if placed in thin compacted layers, was of concern (Moiseev et al., 1988). This method of construction has also been used for a few small- and medium-sized embankment dams in northern China. However, the quality control is not as good as for the compacted "dense" grade asphalt concrete method building the core in compacted layers, and the bitumen content is much higher (about double). To the best of the authors' knowledge, the "flowing" asphalt method has not been used in dam projects outside Russia and China.

Chapter 2

Hydraulic asphalt concrete

2.1 General material requirements and mix proportions

Hydraulic asphalt concrete (HAC) consists of bitumen, rock aggregates, sand, and filler. If required, small amounts of additives (admixtures) are added to the bitumen to increase the flexibility, ductility, and cracking resistance of the HAC and to improve the interaction at the contact with appurtenant structures.

In general, material requirements and mix designs for HAC used in hydraulic structures have built on the experience gained in road and airfield pavement engineering and practice. However, the application and design criteria for asphalt used in pavements, and for HAC used in dam facings and cores are very different, and this is considered in the design of the water barrier.

In general, sound clean and alkaline granular aggregates are preferable because the bitumen itself is acidic, and affinity (adhesion and bonding) between bitumen and aggregates is important. Anti-strip additives may be added to the bitumen to promote adhesion (Section 3.10). Some of the requirements for aggregates to be used in HAC are relaxed compared to the requirements for aggregates used in pavements because the load and environmental conditions are much more favourable for the dam core than those for road and airfield pavements.

Bitumen is commercially produced with various properties, and the grade (type) should be selected to satisfy the requirements specified for the asphalt core. Road pavements may be rehabilitated or replaced every 10–20 years while the dam core is designed for the service life of the dam, which may be 200 years or more.

The optimum properties for the asphalt concrete to be used in the core depend on the load and environmental conditions and on the deformability of the other embankment materials and the dam foundation. The hydro-mechanical properties required for a core in an embankment dam on a non-uniform and compressible foundation in a seismic region are different from the properties specified for a core in a dam on a stiff rock foundation in a non-seismic region.

As shown in Table 2.1, typical HAC core mixes consist of 6.0%–8.0% bitumen of the total weight of the asphalt concrete mix. The contents of different size mineral materials (aggregates)

Table 2.1 Typical asphalt concrete mix for a core in an embankment dam

	Sieve size (mm)			*Bitumen (percentage of total weight)*
	2.36–19.0	*0.075–2.36*	*0–0.075*	
Percentage of mineral weight	45–60	30–40	10–15	6.0–8.0

DOI: 10.1201/9781315177786-2

are commonly expressed in percentage of mineral weight. The mineral material proportions for HAC core mixes are typically 10%–15% of filler material (mineral particles <0.075 mm), 30%–40% of fine aggregates (0.075–2.36 mm), and 45%–60% of coarse aggregates (2.36–19.0 mm) on a square mesh sieve. The particle size distribution (gradation) should satisfy the modified Fuller gradation curve, within specified limits, as described in Section 3.1.

A bitumen content specified to be 6.0%–8.0% of total weight corresponds to 6.4%–8.7% of mineral weight. Bitumen content is expressed as a percentage of total weight in this book.

Note: Particle size for fines (silt size particles) is less than 0.075 mm in the US ASTM standard but 0.063 mm in the European standard (EN).

2.2 Bitumen for use in HAC

Bitumen is a product refined from crude oil, and the properties must meet the national specifications in the country. The grades of bitumen used for road and airfield pavements may be used, but the bitumen content is generally higher, and softer bitumen grades are used for HAC than for bitumen used in pavements.

As an example, Table 2.2 shows the bitumen specifications given in the Chinese standard for asphalt facings and cores for embankment dams (NEA, 2023). It incorporates the use of ASTM and European EN standards and test methods.

Among many parameters, bitumen grades are characterized by the depth of penetration (measured in 1/10 mm) of a standard needle dropped into a bitumen sample under standardized test conditions and procedures (e.g., ASTM). Penetration is expressed by lower and upper limits specified for each grade of bitumen. The penetration must be measured both before and after heating the bitumen for use in the asphalt concrete core mix. The penetration value after heating

Table 2.2 Chinese standard for bitumen used for asphalt facings and cores for embankment dams (NEA, 2023)

Item	*Test method*	*Grade of bitumen*		
		70	*90*	*110*
Penetration: 25°C, 0.1 mm	ASTM D 5	70±10	90±10	110±10
Ductility: 10°C min	ASTM D 113	20	45	40
Softening point, °C min	ASTM D 36	46	45	43
Dynamic viscosity: 60°C, Pa • s min	ASTM D 2171	180	160	120
Solubility in trichloroethylene, % by mass min	ASTM D 2042	99.5	99.5	99.5
Flashing point, °C min	ASTM D 92	260	245	230
Wax content, % by mass max	EN 12606-1	2.0	2.0	2.0
Specific gravity, 15°C g/cm^3	ASTM D 70	Measured	Measured	Measured
Loss on heating for 5 hours at 163°C				
(a) Change by mass, %	ASTM D 6	±0.8	±0.8	±0.8
(b) Drop in penetration, 25°C, % max	ASTM D 5	39	43	45
(c) Ductility: 10°C min	ASTM D 113	6	8	10

Note: The term ductility in this table refers to the maximum tensile elongation of a bitumen specimen with a length of 30 mm and a neck cross-section of 10 mm × 10 mm tested under specified conditions. This is not the common international definition of ductility for various materials. Section 3.6 of this book refers to ductility of hydraulic asphalt concrete (HAC) as the ability to undergo shear deformation without losing shear resistance after peak strength is reached in a triaxial compression test.

may typically be 20%–30% lower than the value measured before heating. Commonly, it is the value before heating that is presented by the producer in a summary of commercial technical specifications.

Bitumen grades that have been used in asphalt cores in dams vary over a wide range, from penetration values as low as 50 to as high as 300. In the past about 30 years, "softer" bitumen types with higher penetration values have been used than for most of the earlier asphalt core embankment dams (ACEDs) built in the 1960s and the 1970s. The use of bitumen grades with higher penetration makes the asphalt core more flexible and ductile for use at seismic sites and are better suited to tolerate differential foundation settlements and core shear distortions without cracking.

The Eberlaste ACED, built in Austria (1963–1968, see Section 10.1.1), is a pioneering structure founded on a very non-uniform and compressible foundation where significant differential settlements under the dam were to be expected. The designer anticipated problems caused by potential cracking in the core if conventional bitumen with penetration values between 60 and 80 were to be used and specified bitumen with a penetration value of 300.

A study of the behaviour of asphalt concrete in bending and extension by using different grades of bitumen is presented and discussed in Chapter 3. Using softer bitumen allows reducing the asphalt mixing temperatures and the required energy input during asphalt concrete production.

Countries in hot tropical and sub-tropical regions may not produce or use "soft" bitumen in the construction of pavements. Bitumen with high penetration values may therefore be difficult to acquire locally, and importation may be required. However, penetration may be increased by using bitumen softeners; for instance, bitumen B180 can be made of a mixture of 85.5% B85 and 14.5% MB1000. If it is not possible to obtain commercial bitumen with a high penetration value, the asphalt core design may have to be modified, or the bitumen content increased to partly compensate for not having bitumen with higher penetration. With increasing bitumen content, viscous creep will be more pronounced and must be considered in the design.

2.3 Aggregates in the HAC mix

Aggregates for asphalt core construction should be made of sound and property-stable rock. The aggregates should be clean, have suitable particle shapes, and have a low flakiness index (FI) with few needle-and-flake-shaped particles. The FI of the aggregate particles is defined as the percentage by weight of aggregate particles whose least dimension is less than 0.6 of the mean dimensions of the aggregates (ASTM). The FI, which at first may appear to be a relatively insignificant parameter in the mix design, proves to be important. It only applies to aggregate particles having a size larger than 6.3 mm and not to finer particles. In road pavement design, the FI is normally required to be less than 15% and not to exceed 25%. As pointed out and discussed in Section 4.4, the flakiness of the aggregates has a significant effect on the stress-strain behaviour of HAC and on the minimum air void content (air porosity) that may be achieved during compaction of the asphalt mix. Flakiness influences (promotes) aggregate fracturing during compaction, and a high FI value requires a higher bitumen content to obtain a workable asphalt concrete mix. Crushed aggregates with needle-and-flake-shaped particles make compaction difficult and can result in less ductile and more brittle behaviour of the compacted asphalt concrete.

Sound aggregates can form stable aggregate skeleton structures to withstand high intergranular stresses. Vibratory rollers are used to compact the hot asphalt during field construction, and the aggregates should be strong enough to undergo such compaction without significant fracturing and breakage of particles.

HAC is designed with more free bitumen between aggregate particles than asphalt used in pavements, and there is less contact between aggregate particles. Aggregate strength as such has little effect on HAC stress-strain behaviour unless intergranular stresses between aggregate particles in a high dam become so large that significant particle breakage may occur. Once in place, aggregates in an asphalt core are not subjected to as severe abrasion as occurs in road pavements. Therefore, the Los Angeles (LA) abrasion test requirements for pavements are unnecessarily stringent when applied to aggregates in a dam core mix. The core in an embankment dam has very favourable environmental conditions with virtually no weathering effects, almost constant temperature during dam operation, and no concentrated static and dynamic loads such as those on pavements. However, the aggregates are heated to about 170°C during HAC production, and heat resistance of the aggregates without signs of deterioration is required.

An example of an unusual case was encountered during the investigation of the suitability of aggregates for the Honggeer ACED built in Inner Mongolia. The aggregates were found to contain sulphur dioxide impurities. When the hot asphalt was placed and compacted, the aggregates produced vapour, and it was very difficult to compact the HAC to meet the specified requirement of maximum 3% air porosity. A special compaction technique had to be developed for the project.

Another and much more common example is from the quarry for the Chimney Hollow ACED in Colorado, USA. The granitic rock contained significant amounts of pegmatite in some regions of the quarry that was not considered suitable for use in the asphalt concrete mix. It was therefore stated in the technical specifications for the project that only areas in the quarry with <5% pegmatite could be used for aggregate production.

The surface of the aggregate particles should have a texture that promotes satisfactory bonding with the bitumen, and the aggregates should possess good adhesion (affinity) to the bitumen. As the rock in the quarry for Chimney Hollow was granitic and therefore acidic, it was assumed that the bonding between bitumen and aggregate would be poor and that an additive would have to be used to promote bonding. However, standardized ASTM bonding tests showed satisfactory bonding and that no additive was required. The complex issue of bonding between bitumen and aggregate is further discussed in Section 3.10.1.

For the sake of improving aggregate gradation and for facilitating workability and asphalt compaction, natural, well-graded sands (with sub-rounded rather than angular particles) are often used as part of or all the fine aggregates in the asphalt concrete mix. However, natural sands, especially those deposited in muddy rivers, are not as clean as crushed aggregates. They must therefore be washed to meet the specified requirements; otherwise, the clay content will cause reduced bonding strength between the sand and bitumen and may result in lowering the strength, decreasing the cracking resistance, and increasing the permeability of the asphalt concrete. Limited clay content in natural sand may be accepted as most of the clay particles can be sucked out of the sand when the sand is heated in the heat drum (Figure 6.1). The clay content should be measured and controlled for the coarse and fine aggregates taken from the hot silo in the asphalt production plant. The clay content must be less than 0.5% in the coarse aggregates and less than 2.0% in the fine aggregates.

2.3.1 *Summary of standard requirements for HAC aggregates*

The requirements for the coarse aggregates (particle sizes between 2.36 and 19.0 mm, on a square mesh sieve) for an asphalt core mix are:

a fresh, sound, and no property changes when being heated; water absorption less than 3% (related to aggregate porosity),
b clean with clay content less than 0.5%,
c the FI should be less than 35% [the main purpose of this requirement for aggregates to be used in road asphalt pavements is to avoid that the flaky particles break down under heavy loads. However, HAC consists of more fine aggregates, more filler, and a higher bitumen content than road asphalt. The layers of asphalt in the dam core are compacted by a much lighter vibratory roller than used for typical pavements, and once the asphalt core is in place, it will not be subject to high point loads. Therefore, the requirement on the FI for coarse aggregates used in an asphalt core could be relaxed (i.e., limit increased) compared to that for pavements, and a limit higher than 35% could be recommended. However, a high FI conflicts with achieving low air void content during compaction and achieving high ductility of the HAC. This is further discussed in Section 4.4],
d good durability: the loss by mass must be less than 12% after five cycles of dry and wet disintegration by crystallization,
e abrasion loss must be less than 45%. For road pavement design, the abrasion loss must be less than 30% after 500 revolutions in the standard LA abrasion test. ICOLD (2018) recommends that it should be less than 40% for the coarse aggregates in a dam core as the core is subjected to much less severe abrasion than a road pavement. The authors recommend that the limit is further increased to 45%. The aggregate gradations in the compacted asphalt layer must be checked and controlled,
f good adhesion: after hot mixing, the entire aggregate surface must be covered with bitumen and good physical and/or chemical adhesion (bonding) between aggregate and bitumen must be achieved. Commonly, the adhesion is evaluated by the standard boiling water method. The requirement [Grade 4, NEA (2019)] is that the remaining area of bitumen covering the aggregate surface should be larger than 90% after 3 minutes in boiling water or larger than 85% after 10 minutes in boiling water (ASTM, 2005). Alternatively, one may evaluate adhesion by using the rolling bottle test criteria described in the European Standard (2020a) (see Section 3.10.1),
g the coarse aggregates in the asphalt mix must be well-graded.

The requirements for the fine aggregates (particle size between 0.075 and 2.36 mm) are:

a fresh, sound, and no property changes when being heated,
b clean, no organic substances or other impurities, and clay content less than 2.0%,
c good durability: loss by mass less than 15% after five cycles of dry and wet disintegration by crystallization. The LA abrasion test is not required for aggregate particle sizes smaller than 2.36 mm,
d the fine aggregates must be well-graded.

2.4 Filler in the HAC mix

Filler is the mineral powder (fines and silt size) produced by aggregate crushing and/or grinding, most of which has a particle size less than 0.075 mm by ASTM standard (or 0.063 mm by European standard). In an HAC core mix, the filler material constitutes 10%–15% of mineral weight, and the filler's surface area makes up about 95% of the surface area of all the mineral materials in the mix. Fillers serve a very important function and must have suitable fineness and good adhesion to the bitumen used.

Fillers can consist of ground powders of limestone, dolomite, and carbonate. Portland cement, after being tested to document the required quality, can be used as filler in the asphalt core mix. Fly ash has also been used as filler material to make successful asphalt concrete mixes. In general, not more than 50% of the filler should come from the fines produced during the aggregate crushing at site unless the aggregate consists of pure limestone, dolomite, or carbonate. The remaining amount of filler should be obtained from a cement plant or other quality sources.

As an example, for the Huangjinping ACED in Sichuan Province, 100% of the filler came from the fines produced during the crushing of dolomite at site. The dolomite aggregates used showed an LA abrasion loss of 27%, a value that met the requirement of maximum 30% for road asphalt. The filler content in the crushed fine aggregates turned out to be higher than that required for the asphalt core mix design. Therefore, a special technique was used to modify the asphalt plant to control the filler content in the asphalt concrete mix.

When acidic aggregates are used, it is inevitable that the fine aggregates contain some acidic powder. The acidic filler content should be restricted to less than 50% of the total weight of the filler in the mix. Simultaneously, an anti-strip additive may be recommended to improve the bitumen-aggregate adhesion. For the Nuer ACED (80 m, 2017) in China, acidic granite aggregates were used in the asphalt core mix, and it was decided that an additive should be used to promote adhesion. However, as noted and discussed in Section 3.10.1, an anti-strip additive may not lead to much improvement of the bitumen-aggregate adhesion for the HAC in a dam core, while it has proved very effective in road pavements.

In summary, the requirements for fillers for HAC are:

a well-graded,
b water content less than 0.5%,
c hydrophilic coefficient less than 1.0 (indirectly testing for adhesion to bitumen),
d no clay or organic substances or other impurities.

Chapter 3

Asphalt concrete mix design and suitability testing

3.1 Aggregate gradation

In the years around 1910, Fuller and colleagues developed and proposed an aggregate gradation curve to achieve optimum aggregate packing of the form:

$$P_i = 100\left(\frac{d_i}{D}\right)^{0.5} \tag{3.1}$$

where:

P_i = percentage by weight of particles smaller than the equivalent grain size dimension d_i,
d_i = size of sieve mesh (mm),
D = maximum grain size of aggregates (mm).

Later, it was proposed that Fuller's index of aggregate gradation (0.5 in Eq. 3.1) be taken as a variable (n) to make the aggregate gradation curve more general (Ding, 1982). Theoretical and experimental studies in road pavement engineering suggested that maximum density may be achieved by taking the index of aggregate gradation in the Fuller equation to be 0.45. An index of aggregate gradation that gives maximum density was commonly assumed to give the highest compressive strength and can sustain the highest shear stresses with the least particle breakage.

For a given maximum grain size D, the proportions among coarse aggregates, fine aggregates, and filler are fixed by applying the Fuller equation. ICOLD (2018) recommends that the mineral aggregate composition for an asphalt concrete mix to be used in a dam core shall comply with Fuller's gradation curve with an index of aggregate gradation $n = 0.45$ and improved with a fine grain component (i.e., filler material with a grain size smaller than 0.075 mm).

The mineral grain composition recommended in the Chinese specifications has been modified somewhat compared to the ICOLD (2018) recommendations and the basic Fuller gradation curve. Ding (1982) and Ding et al. (1991) proposed the following equation that has been adopted in the Chinese test code for the design of asphalt concrete facings and cores for embankment dams (NEA, 2023):

$$P_i = F + (100 - F)\frac{d_i^n - d_{0.075}^n}{D^n - d_{0.075}^n} \tag{3.2}$$

where:

F = filler content with a particle size less than 0.075 mm (in percentage),
$d_{0.075}$ = size of sieve mesh, 0.075 mm.

DOI: 10.1201/9781315177786-3

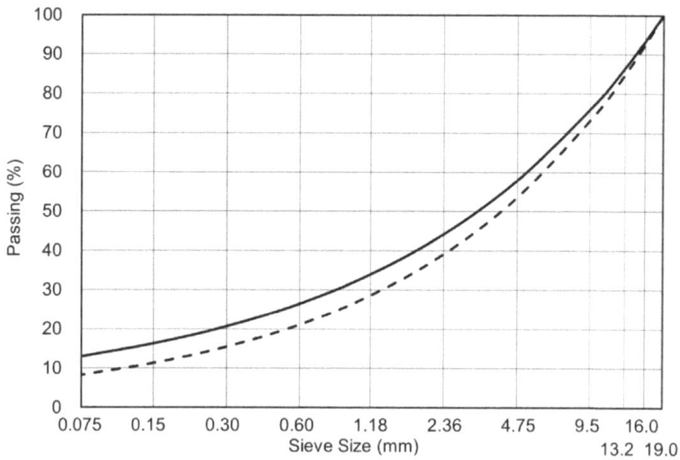

Figure 3.1 Fuller gradation (dashed line, $n=0.45$, $F=8.3\%$) and modified Fuller gradation (solid line, $n=0.41$, $F=13.0\%$) curves with a maximum grain size of 19.0 mm, metric sieves. The base of the logarithm for a sieve size in mm is 2.

The advantage of using Eq. (3.2) is that the proportions among coarse aggregates, fine aggregates, and filler can be independently adjusted depending on the different mineral materials available to determine optimal mineral particle size distribution. The commonly used square opening metric sieves for asphalt concrete aggregates are 19.0, 16.0, 13.2, 9.5, 4.75, 2.36, 1.18, 0.60, 0.30, 0.15, and 0.075 mm. Figure 3.1 shows a Fuller gradation curve and a modified Fuller gradation curve with $D=19.0$ mm (max. grain size), $n=0.41$, and $F=13.0\%$.

The aggregate gradation curve (Eq. 3.2) is designed to make the coarse and fine aggregates form the densest aggregate skeleton, called dense grade.

Due to the differences in aggregate shapes and textures among various lithologies and processing technologies, specifying only one aggregate gradation exponent may not yield the maximum density for a given mineral material. The compressive packing model (CPM) is therefore used to optimize asphalt concrete mixes for pavements (Baghaee et al., 2018), and it may be advantageous in future practice to use the CPM model to further improve the aggregate gradation design also for hydraulic asphalt concrete (HAC).

3.2 Bitumen content and filler content

The bitumen content and filler content should be designed to make the volume of bitumen and filler theoretically fill all the voids within the aggregate skeleton of coarse and fine aggregates. Experience shows that the bitumen content in satisfactory HAC mixes in dam cores is usually between 6.0% and 8.0% of total weight and the filler content between 10% and 15% of mineral weight. The ratio of filler to bitumen content by weight is usually about 1.8 (Zhang et al., 2008).

3.3 General requirements for HAC properties

For HAC to be "impervious", the bitumen content should be 0.2%–0.3% points higher than that required to just fill the voids within the aggregate skeleton. This means that the asphalt mix is slightly "supersaturated" with a small amount of "free" bitumen between the aggregate particles. The free bitumen reduces the strength of the asphalt concrete but makes it more flexible and ductile and close to impervious. Increased ductility implies reduced brittleness and reduced strain-weakening (strain-softening) after peak shear strength has been reached in a shear test, for instance in a triaxial compression test as described in Section 3.7.

Research and experience show that the air void content (air porosity) of the asphalt concrete in the core must not be more than 3.0% to ensure that the dam core is virtually impervious (permeability coefficient approximately 10^{-10} cm/s). The asphalt concrete in the core should be able to undergo relatively large shear strains without significant volume expansion (shear dilation), which can lead to an increase in permeability. More importantly, and as discussed in Section 5.1, large shear strains and volumetric expansion can lead to the development of an interconnected network of fissures that can cause a very significant increase in permeability. A softer grade of bitumen and/or a higher bitumen content should be used to make the asphalt core better able to accommodate large distortional shear strains without excessive dilation and possible fissuring. Triaxial compression tests clearly show that by increasing the bitumen content and by increasing the confining stress, the degree of shear dilation decreases (Figure 3.2). The triaxial

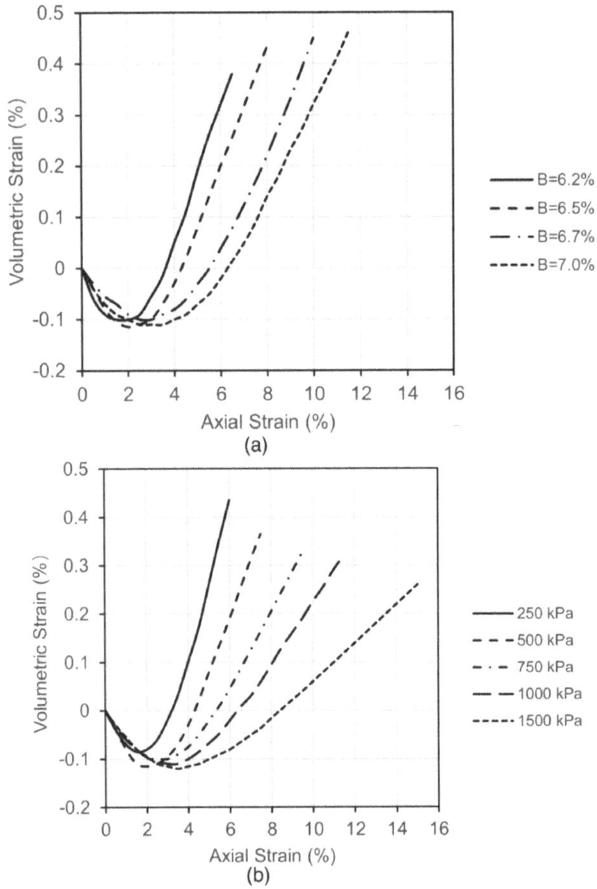

(a) Triaxial compression tests with lateral confining stress 500 kPa and various bitumen contents (B).
(b) Triaxial compression tests with bitumen content 6.5% and various lateral confining stresses.

Figure 3.2 Volumetric strain as function of (a) bitumen content and (b) confining stress in triaxial compression tests. Positive volumetric strain means volumetric expansion (dilation).

compression tests referred to in the figure were conducted at an axial strain rate of 0.1%/minute and at a temperature of 10°C. When the applied axial strain rate is reduced, the degree of shear dilation is reduced.

The asphalt concrete to be used in the dam core must be workable and can be compacted to the required density. As discussed in Chapter 4, compaction methods traditionally used in the laboratory are significantly different from the roller compaction commonly used in the field, and the stress-strain behaviour of laboratory compacted specimens may be very different from the stress-strain behaviour of field-compacted specimens of the same asphalt mix. The asphalt mix design based on laboratory-prepared specimens must always be validated by performing the required field test section investigations described in Section 6.6 using aggregates crushed by the field crushers of the rock to be used in the project. As a result of the field test section investigations, the preliminary asphalt mix design suggested by the laboratory investigations may have to be modified to ensure that the HAC has the characteristics required for core construction.

3.4 Marshall standard method of mix design and test specimen preparation

Traditionally, the Marshall laboratory methods and test procedures developed for pavements have been applied when making the asphalt concrete mix for the dam core. Laboratory specimens are prepared in a standardized cylindrical steel mould with an inner diameter of 101.6 mm, and the asphalt specimen after compaction has a height of 63.5 mm.

The aggregates are first dried at a temperature of 110°C and then sieved in six particle size ranges: 19.0–16.0, 16.0–13.2, 13.2–9.5, 9.5–4.75, 4.75–2.36, and ≤2.36 mm. The dry aggregates in each particle range and the added filler are weighed in accordance with the specified weight proportions, and they are preheated for 4 hours at 170°C. The bitumen is preheated for 2 hours at about 160°C, and the asphalt concrete specimens are prepared and compacted as described below. The temperatures given above are suitable for mixes with bitumen penetration values of 70–90 (called grades B 70 and B 90, respectively). The specified temperatures depend on the grade of bitumen used and are somewhat lower for softer bitumen. The hot aggregates and the bitumen are placed in a mixer and mixed for 1.5–3.0 minutes to make all the aggregate and filler particles completely coated with bitumen. At a temperature between 155°C and 165°C, the mixture is placed and compacted in the preheated cylindrical mould according to the Marshall standard method and procedure. For bitumen with a high penetration value, for instance B180, a somewhat lower (15°C lower) temperature should be used. Exact temperatures for mixing and compacting an asphalt concrete mix mainly depend on the dynamic viscosity of the bitumen and on aggregate characteristics (Table 2.2). The temperature of the aggregates is usually required to be not more than 20°C above the temperature of the bitumen to prevent bitumen from "over-ageing" (see Section 5.3).

The standard Marshall hammer (tamper) for specimen compaction in the cylindrical mould has a flat circular tamping surface with a diameter of 98.5 mm, weighs 4.5 kg, and the standard height of the hammer drop is 457 mm. The hot asphalt concrete is compacted by 20 blows on the top and 20 blows on the bottom of the specimen. In special situations, another predetermined number of compaction blows may be specified. The hydro-mechanical properties of the compacted mix depend on the number of blows used during compaction. After compaction and cooling, the specimen is taken out of the mould, cooled for 24 hours at ambient temperature, weighed, and the air void content (air porosity) is computed.

For the asphalt core mix design with the Marshall standard method and procedure, it is normal to prepare specimens with three different aggregate distribution curves, three different filler contents, and four different bitumen contents in the range considered relevant for most dam cores, i.e., 6.0%–8.0% of total weight of asphalt mix. If crushed, coarse aggregates are mixed with natural gravel, and two or three different proportions of crushed aggregate and natural gravel are tested. If natural sand is used as fine aggregates, the gradation should be examined, and three proportions of natural sand and crushed sand should be tested. Natural sand with naturally subrounded particles, rather than crushed angular sand, is often used to make the mix more workable. It should be noted that natural sand has a significantly lower angle of shearing resistance (friction angle) than crushed, angular sand and reduces the shear strength and compressive strength of the asphalt concrete mix.

The relationships between air void content (air porosity) and aggregate gradation, filler content, bitumen content, and proportions of natural and crushed aggregates should be investigated. The asphalt core mix to be used in a project should, after the prescribed, standardized laboratory compaction, be required to give the Marshall standard specimen an air void content less than 2.0%. Experience shows that if the Marshall standard specimen after laboratory compaction has an air void content lower than 2%, standard roller compaction in the field will normally achieve an air void content lower than 3%. In the dam core, the air void content (air porosity) is usually required to be not more than 3%, which gives a virtually impervious core.

However, if the flakiness index (FI) of the aggregates larger than 6.3 mm (see Sections 2.3 and 4.5) is higher than 20%, it may be difficult to achieve an air void content for the compacted standard Marshall specimen as low as 2%. Rather than using the standard Marshall mould and method of compaction, a mould and method better simulating field compaction (SFC) should be used as discussed in Chapter 4. For the asphalt mix to be acceptable, it should be required that the air void content of the SFC specimen after laboratory compaction should be less than 2%.

Standard tests on the 63.5 mm high and 101.6 mm diameter Marshall specimens are used to determine the standard Marshall "stability and flow" values for the trial mix (Asphalt Institute, 1979). In asphalt pavement design, these parameter values are considered essential in evaluating the suitability of a proposed mix for pavement construction. However, in the design of an asphalt core for an embankment dam, the technical requirements are different, and more emphasis is placed on the results of triaxial stress-strain-strength tests and tension and bending tests as described in Sections 3.7, 3.8, and 3.9.

3.5 Permeability (hydraulic conductivity) of compacted HAC

Figure 3.3 shows the principle of the test equipment that is used to investigate the permeability of compacted asphalt specimens under high constant water pressure.

Figure 3.4 shows the test results of permeability versus air porosity of 40 specimens with air porosity between 2% and 11% performed by Kjærnsli et al. (1966) using test equipment similar to that shown in Figure 3.3. There is some scatter in the test results but later tests by other researchers have confirmed the early findings. ICOLD (2018) concludes that the relationship between air void content and permeability for specimens prepared and compacted by the Marshall method is documented through many studies. For asphalt concrete to be virtually impervious and the dam core to be watertight, the air void content (air porosity) should be maximum 3%. The permeability coefficient is then not more than about 1×10^{-10} cm/s for a specimen prepared and compacted by the standard Marshall method (Section 3.4). In standard practice,

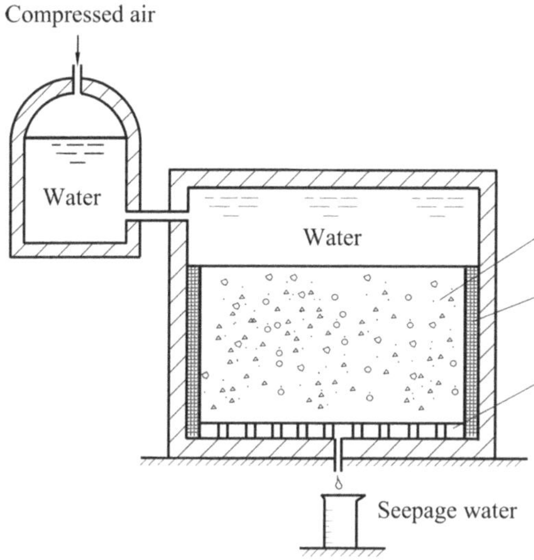

Figure 3.3 Permeability test equipment at Xi'an University of Technology using high hydraulic head for dense-grade hydraulic asphalt concrete (HAC) specimens.

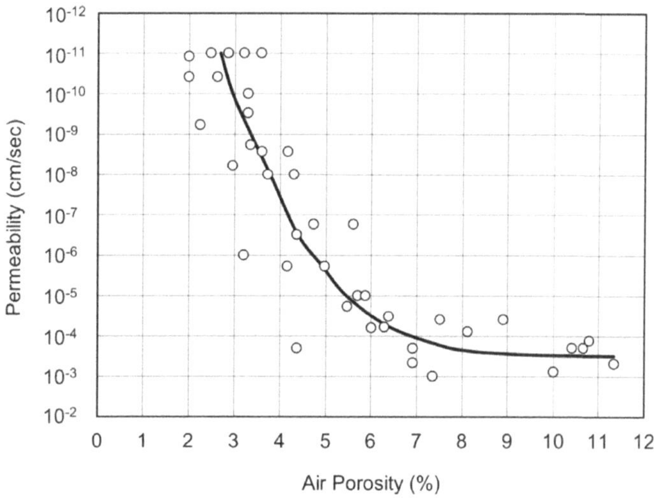

Figure 3.4 Permeability versus air void content (air porosity) of asphalt concrete (Kjærnsli et al., 1966).

it is normally accepted that permeability tests are not required during core construction, but frequent air void content measurements are considered essential to control that the air porosity after specified compaction is maximum 3%. If that requirement is not satisfied, the asphalt mix must be adjusted to satisfy the air porosity criterion in the subsequent core construction (see Section 6.6). Some of the already placed and compacted asphalt concrete in the core must be removed and replaced or the already placed portion of the core must be made watertight by other remedial measures.

It should be noted that the ICOLD conclusion cited above is only valid for asphalt mixes designed by following the standard Marshall method and procedure (Section 3.4). If an asphalt mix is designed with less fine aggregates, less filler content, and less bitumen than that recommended by the standard Marshall mix design, the permeability could be much larger than 10^{-10} cm/s due to more interconnected fissures and voids in the specimen. This may be the case even if the asphalt specimen is compacted to meet the specified air void content of no more than 3%.

3.6 Differences between aggregates crushed in the laboratory and in the field

The early laboratory suitability tests on the proposed asphalt concrete mix to be used in the project are usually run with aggregates crushed in the laboratory. The aggregates crushed by the field crusher will be different from those used in the laboratory tests, and the test results for specimens prepared with field and laboratory aggregates, respectively, may be different. Experience shows that the hydro-mechanical test results from specimens produced in the laboratory may be somewhat misleading. The final suitability tests must be run on specimens with aggregates crushed by the field crushers for the project and on the rock type to be crushed for the asphalt concrete production. The FI, which has significant effects on the HAC behaviour, may be quite different for aggregates crushed in the laboratory rather than by the field crusher and for rock types of somewhat different origin.

3.7 Triaxial testing to investigate stress-strain-strength behaviour

Triaxial tests are normally performed on asphalt specimens of 100 mm in diameter and 200 mm in height, and the specimens are mounted in the triaxial pressure chamber (Figure 3.5). The tests are carried out at a specified temperature, usually between 5°C and 20°C. The chamber is filled with deaired water, and the pressure in the chamber may be increased by compressed air as shown in Figure 3.5. Triaxial tests may be performed in various ways by increasing or decreasing the axial stress and the radial confining stress on the specimen, i.e., by specifying different stress paths. In conventional triaxial compression tests, the axial stress or axial strain is gradually increased while the radial confining stress is kept constant.

The asphalt specimen is usually enclosed in a thin, impervious rubber membrane during the triaxial testing, as is done in the testing of soils in geotechnical engineering. The membrane prevents water in the chamber from penetrating into the asphalt specimen when fissures and surface cracks open up in the asphalt specimen at high shear stress levels. This gives the possibility of studying the complete stress-strain curve and provides for a well-defined stress condition to be analysed. Wang and Høeg (2002) reported the results from tests with and without the use of rubber membrane and confirmed that until the specimen starts opening cracks, the recorded stress-strain curves are the same for the specimen whether it is tested with or without rubber membrane.

The magnitude of applied radial confining stress on the laboratory specimen depends on the height of the dam and the reservoir water level, the properties of the embankment materials, and the lateral stresses induced by the compaction of transition Zone 2 on both sides of the core. The stress-strain behaviour of the triaxial test specimen depends on the level of applied radial

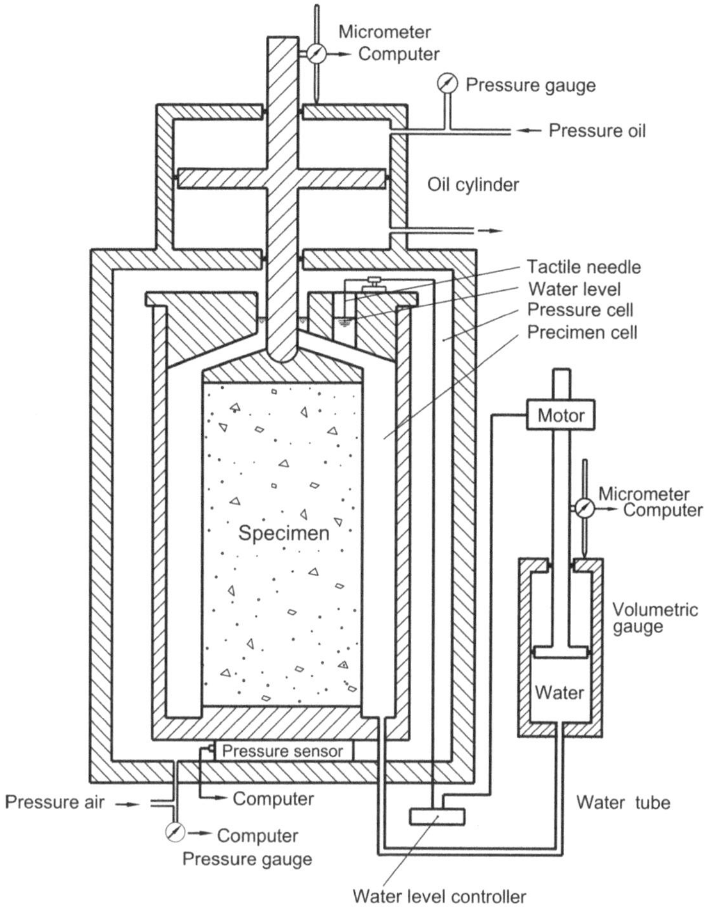

Figure 3.5 Triaxial chamber at Xi'an University of Technology for testing of asphalt concrete.

confining stress. Most triaxial tests performed for asphalt concrete have been done by imposing a constant axial displacement rate (i.e., strain-controlled test). The imposed axial strain rate and test temperature must be specified in the laboratory report presenting the test results that strongly depend on strain rate (loading rate) and temperature during testing.

Axial strain rates for asphalt concrete triaxial tests published in the literature have usually been about 1%/hour (0.017%/minute). Most published data are for tests with approximately this strain rate and a radial confining stress in the range 200–1,200 kPa.

The magnitudes of axial strain, axial stress, and volumetric strain (volume reduction or volume expansion) are measured and recorded throughout the test. Typical stress-strain results are shown in Figure 3.6 for two tests where the axial stress (σ_1) is increased and the radial confining stress (σ_3) is kept constant during the test. In strain-controlled tests, the stress-strain curve can also be recorded for strains after the peak strength is reached and when there is a loss of shear resistance with a further increase in shear strain [i.e., strain-weakening (strain-softening) behaviour of the specimen].

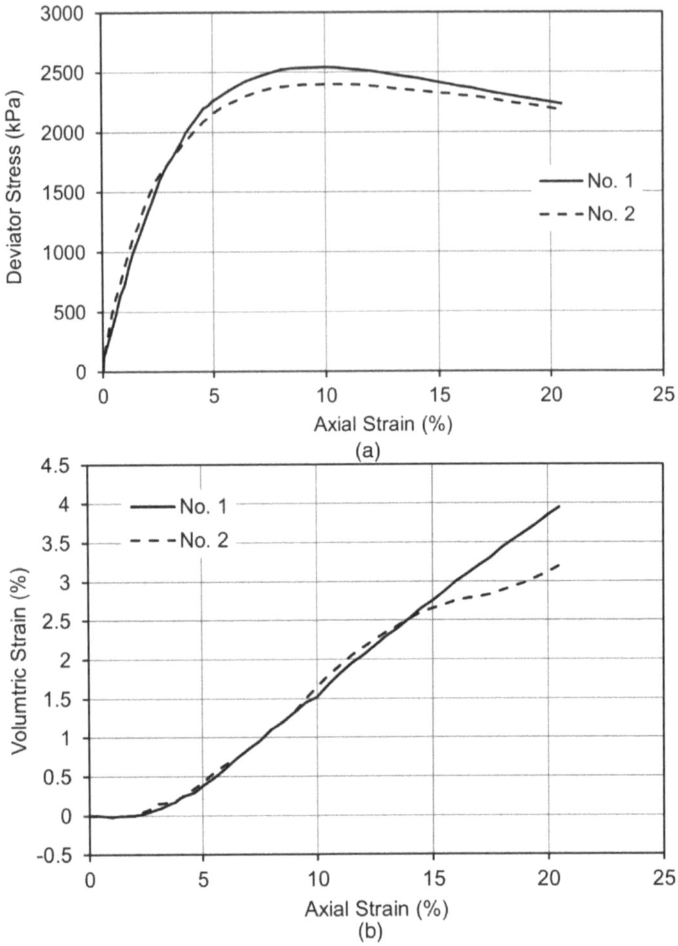

Figure 3.6 Deviator stress (a) and volumetric strain (b) versus axial strain at a radial confining stress of 500 kPa for two asphalt specimens prepared of the same mix and compacted the same way in the laboratory (specimen dimensions: 100 mm in diameter and 200 mm in height, tested at an axial strain rate of 0.033%/minute at a temperature of 10°C).

Figure 3.6 shows typical results for two triaxial compression tests performed on the same asphalt concrete mix at 10°C with an axial strain rate of 2%/hour (0.033%/minute). Both specimens were enclosed in impervious rubber membranes during the triaxial tests. During preparation, the specimens were compacted in four layers, each about 60 mm thick before compaction, to the same air void content (air porosity). The initial confining stress for both specimens was 500 kPa and was kept constant during the tests. The resulting stress-strain curves are very similar for the two specimens except for some differences in measured volumetric strains at large axial strains as the specimens approached failure and significant cracking started to occur. Both tests gave an initial compression modulus (Young's modulus) of about 100 MPa for the given confining stress, temperature, and strain rate and a maximum deviator stress (peak strength) of about 2,500 kPa at 10% axial strain. The initial compression modulus is here defined as the secant modulus from 0% to 1% axial strain.

After maximum deviator stress is reached, the stress-strain curves dip down slightly, but the behaviour is classified as ductile with minor strain-weakening (strain-softening). The weakening is only down to a deviator stress of 2,200 kPa at 20% axial strain, which is 10% lower than at the maximum deviator stress. The volumetric expansion is 1.5% at maximum deviator stress and about 2.5% at 15% axial strain.

Figure 3.7 shows the triaxial test results for several specimens drilled out of the compacted asphalt core in the 174-m high Quxue asphalt core embankment dam (ACED) that is presented

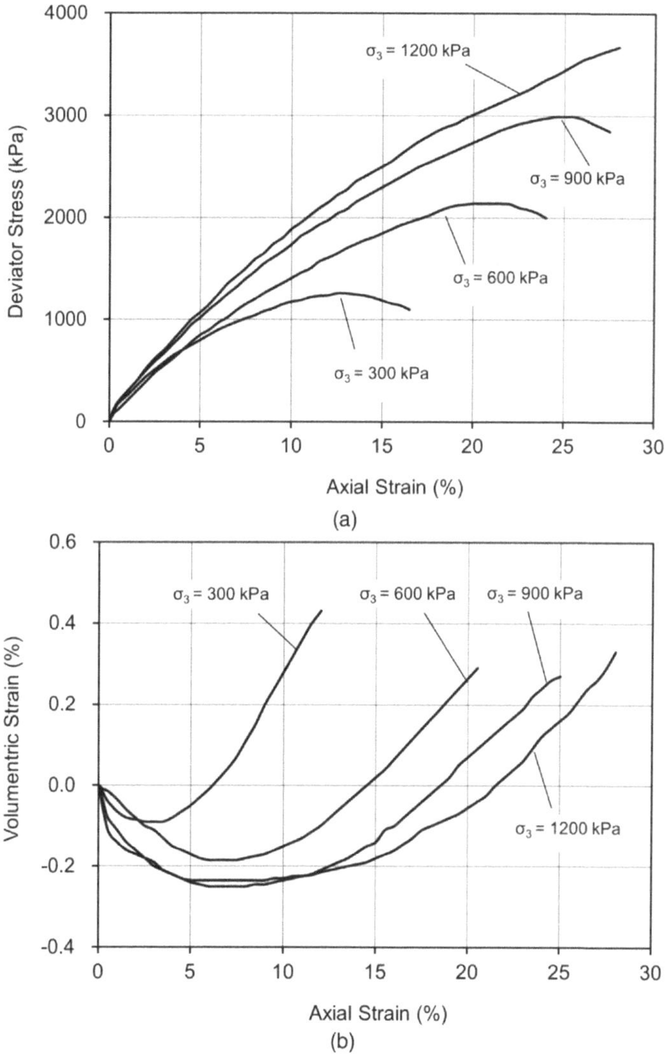

Figure 3.7 Deviator stress (a) and volumetric strain (b) vs. axial strain for 4 (four) different radial confining stress levels for asphalt core samples drilled out of the roller-compacted core of the 174-m high Quxue ACED. Each curve represents the average results of 4 (four) triaxial tests at each level of confining stress (specimen dimensions: 100 mm in diameter and 200 mm in height, tested at an axial strain rate of 0.1%/minute at a temperature of 15°C).

in more detail in the case study in Section 10.4. At the axial strain rate of 0.1%/minute and test temperature 15°C, the initial compression modulus for the roller-compacted asphalt in the Quxue ACED core is found to be between 15 and 20 MPa. From Figure 3.7, it may be observed that the maximum deviator stress levels are about 1,250, 2,100, 3,000, and more than 3,500 kPa for confining stress levels of 300, 600, 900, and 1,200 kPa, respectively. For the low confining stress level (σ_3=300 kPa), the tendency to volume expansion (dilation) started after an axial strain of about 3%, while for σ_3 =1,200 kPa, the tendency to volume expansion did not start until an axial strain of about 7%.

It should be noted that the initial compression modulus (taken as the secant modulus up to the axial strain of 1.0%) and the peak strength are strongly dependent on the imposed strain rate. This is further discussed in Chapter 7.

Triaxial testing of the specimens shows whether the proposed asphalt concrete mix exhibits ductile stress-strain behaviour in the core after maximum deviator stress is reached. Ductile behaviour is very advantageous for asphalt concrete to be used in a dam core. The triaxial testing performed during the early mix design is mainly qualitative and is used to compare the general behaviour of various asphalt mixes to select a suitable mix for a given asphalt core project (suitability testing). Subsequent triaxial testing on the final mix may be carried out to determine stress-strain parameters to be used in numerical stress-deformation analyses.

To simulate realistic field loading conditions in the laboratory, long-term triaxial creep tests are required, and this is discussed in Chapter 7.

During preliminary design and suitability testing of the asphalt concrete mix, test specimens have to be prepared and compacted in the laboratory. At a later stage in the project, specimens may be drilled out of the required test section (see Section 6.6) and that provides more realistic stress-strain curves and parameters as discussed in Chapter 4. During dam construction quality control, triaxial specimens may be drilled out of the constructed core to determine stress-strain-strength properties. However, by that time in the project, many design decisions have already had to be made.

As discussed in Chapter 4, there may be significant differences in stress-strain behaviour of specimens prepared and compacted in the laboratory and specimens drilled out of (extracted from) the asphalt core compacted by vibratory rollers in the field.

3.8 Behaviour in direct tension and in bending

The ACED dam core is predominantly in compression, but depending on the foundation conditions, tensile stresses and extensional strains may develop in the longitudinal direction. Impounding causes bending in the core wall in the upstream-downstream direction, and local stress and strain concentrations may potentially cause cracking in the core. Furthermore, earthquake loading causes dynamic and residual tension that may lead to cracking. Therefore, the behaviour of asphalt concrete in tension and bending must be investigated by appropriate testing. The test results are used in core design and in the evaluation of field performance and safety margins, as discussed in Chapters 8 and 9.

The specimen size for direct tension tests in the laboratory is typically 40×40 mm^2 in cross section and 220 mm long with an effective length between the tension grips of 160 mm. Figure 3.8 shows multi-function test equipment used for direct tension, stress relaxation, and crack-opening tests. The direct tension test is usually performed at constant temperature and with a constant axial extension strain rate. In Figure 3.8, a specific displacement rate (strain rate) is achieved by a servo motor. The tension force and the displacements are measured by

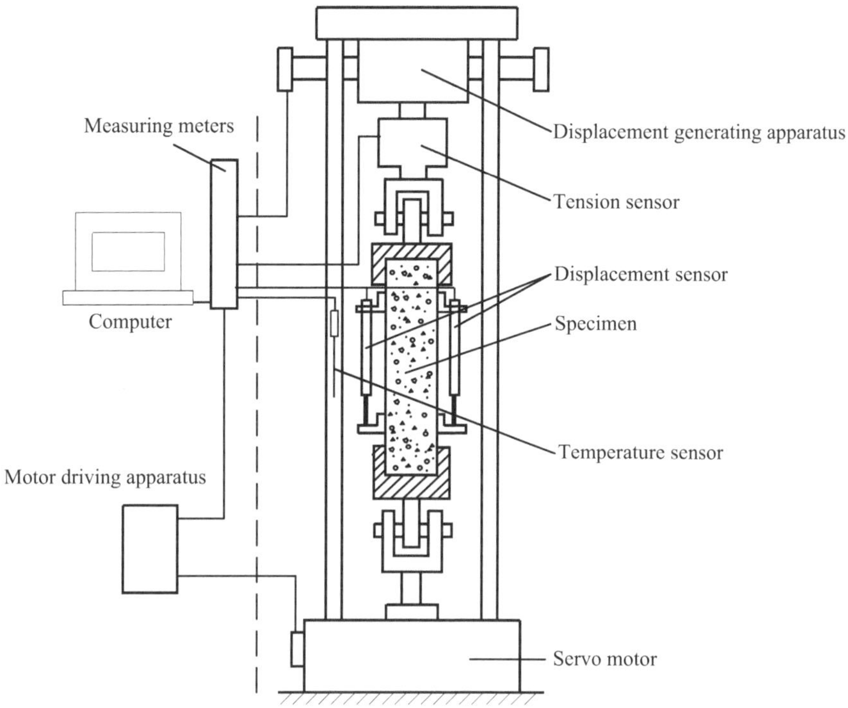

Figure 3.8 Multi-function test equipment at Xi'an University of Technology for direct tension, stress relaxation, and crack-opening tests for asphalt concrete.

the tension sensor and the two displacement sensors, respectively. The axial force and displacements are recorded during the testing until the specimen fractures and fails.

Figure 3.9 shows a sketch of the beam bending tests. The specimen is a simply supported beam of cross section typically 35 mm (wide), 40 mm (high), and 220 mm (long) with an effective span length between supports of 200 mm.

In single-point beam loading, the load is applied at the midpoint of the beam span and symmetrically about the midpoint for two-point loading. Tests are performed with a specified deflection rate and temperature. The magnitudes of the maximum bending stress and bending strain in single-point beam loading, assuming linearly elastic, isotropic material behaviour, are:

$$\sigma = \frac{3LP}{2bh^2} \tag{3.3}$$

$$\varepsilon = \frac{6hf}{L^2}100\% \tag{3.4}$$

where:

σ = maximum bending stress (MPa),
b = width of cross section of the specimen (mm),
h = depth of cross section of the specimen (mm),
L = span length of the specimen (mm),
P = load applied at the midpoint of beam span (N),

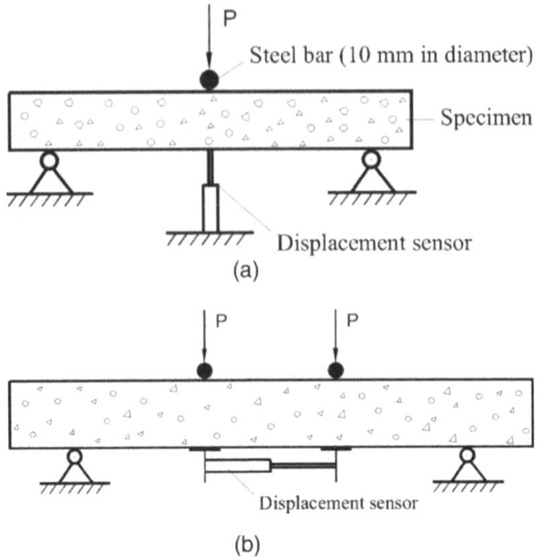

Figure 3.9 Beam bending tests with (a) one-point and (b) two-point loading. Measurements of vertical beam deflection and longitudinal, extensional strain in bottom of beam between loading points.

f = mid-span deflection (mm),
ε = maximum bending strain in beam (%).

The asphalt beam for the two-point loading test is typically $50 \times 50\,mm^2$ in cross section and 400 mm in length with a total effective span between the supports of 300 mm.

In special cases, circular plate bending tests are also performed as described by Høeg (1993). The circular plate was 50 mm in thickness and 600 mm in diameter with an effective diameter of 500 mm. The plate bending test is primarily used in connection with the design and performance of upstream facings and linings.

The results from beam and plate bending tests may be used to quantify the magnitude of tolerable bending strains before cracking occurs. During design, the tolerable values should be compared with the bending strains (computed or measured) in the dam core after impoundment and after any additional bending deflections (computed or measured) after earthquake loading, to evaluate whether cracking of the core is likely to occur due to bending.

Figures 3.10 and 3.11 show the direct tension and beam bending test results of asphalt specimens drilled out of the core of the Quxue ACED (see Section 10.7). In the direct tension tests, the tensile strength and strain when cracks developed and lead to specimen failure, were found to be 0.5 MPa and 1.6%, respectively. In the beam bending test, the bending strength is taken as the stress when cracking caused by bending leads to failure of the beam. The back-calculated strength and strain values were 1.0 MPa and 6%, respectively. Direct tensile strength primarily depends on the properties of the bitumen with filler as the aggregates in the asphalt mainly play an “inactive role”, while bending behaviour is a composite action of compression and tension behaviour in the asphalt concrete specimen. This may explain most of the difference in the magnitudes of strain to cause cracking in direct tension and beam bending tests, respectively.

The results for various asphalt mixes tested in the range 0°C–20°C show significant differences in magnitude of the strains at cracking between direct tension and bending tests.

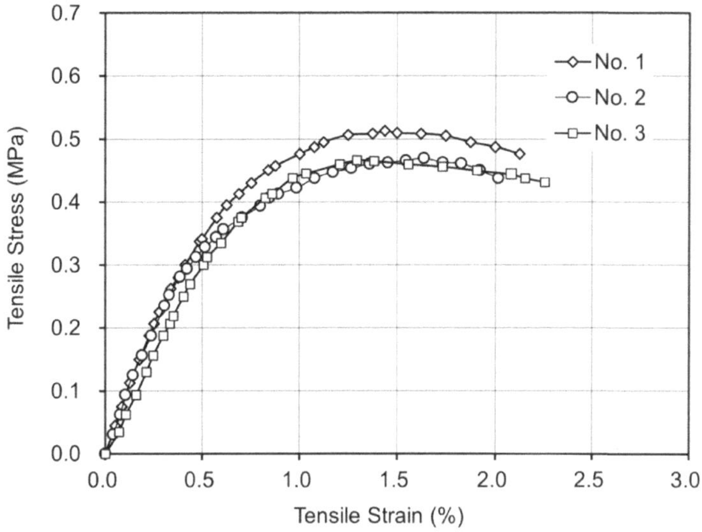

Figure 3.10 Tensile stress vs. tensile strain for three specimens drilled out of the asphalt core in the 174-m high Quxue ACED (tested with an axial strain rate of 1.0%/minute at a temperature of 15°C).

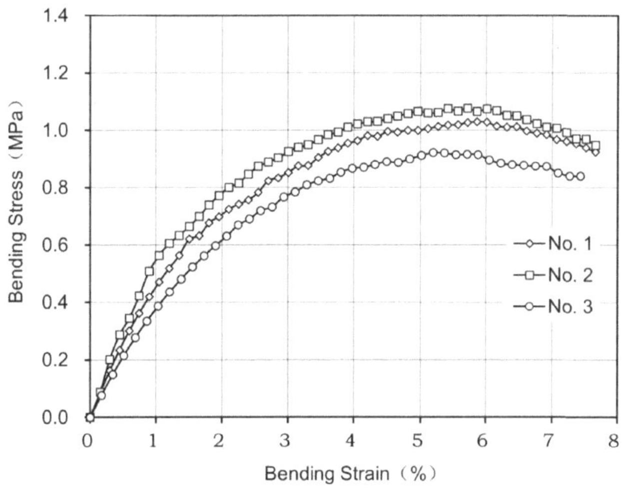

Figure 3.11 Bending stress vs. bending strain for three specimens drilled out of the asphalt core in the 174-m high Quxue Dam (tested with a bending strain rate of 1.0%/minute at a temperature of 15°C in single-point beam loading tests with a beam cross section of $35 \times 40\,mm^2$).

The tensile strain at cracking for the bending tests is commonly 2–3 times higher than that for the direct tension tests. The strain at cracking at 20°C is about 2.5 times higher than that at 0°C (Zhang et al., 2013).

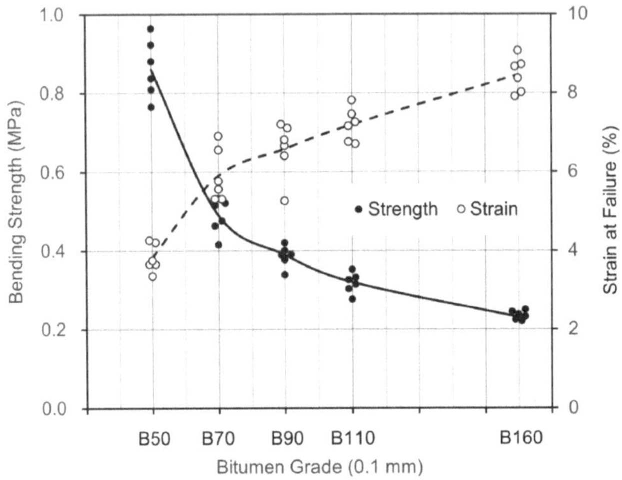

Figure 3.12 Bending strength and bending strain at failure vs. bitumen grade for asphalt concrete specimens tested with a bending strain rate of 1.0%/minute at a temperature of 20°C.

The bitumen grade has pronounced effects on the asphalt concrete compression, tension, and bending behaviour. Bitumen grades B50, B70, B90, B110, and B160 from the Kelamayi refinery in China were selected in a comprehensive comparative study. The penetrations of these grades of bitumen were checked in the project and measured to be 47, 71, 87, 106, and 161 (0.1 mm), respectively. Limestone aggregates and filler were used applying the modified Fuller aggregate gradation in Figure 3.1. The bitumen content was 6.4% of total weight. Figure 3.12 shows the bending stress and bending strain when the strength was reached vs. bitumen grade for the asphalt concrete specimens tested with a deflection rate of 1.67 mm/minute (bending strain rate of 1.0%/minute) at a temperature of 20°C.

Figure 3.12 shows the significant effects of bitumen grade on the bending behaviour of asphalt concrete. The maximum bending stress when the bending cracks opened was 0.49 MPa for asphalt concrete using bitumen B70, while it was only about half of that (0.23 MPa) for asphalt concrete using bitumen B160. The bending strain at failure was 5.9% for asphalt concrete using bitumen B70, while it was significantly higher (8.5%) for asphalt concrete using bitumen B160.

3.9 Splitting (Brazilian) test to determine tensile strength

The Brazilian splitting test (European Standard, 2020b; NEA, 2019) is a method to indirectly determine the tensile strength of HAC. The test is much easier to prepare and to perform than the direct tension and bending tests described in Section 3.8.

A cylindrical asphalt concrete specimen is compacted in the laboratory in the same mould as used for preparing a standard Marshall specimen or a triaxial test specimen, or it is drilled out of the test section or the dam core. The long drilled-out specimen (usually about 450 mm) is cut into slices with thickness about half of the diameter of the drilled core (see Figure 6.17a). The slice is placed between an upper and lower jaw in the testing machine, and a vertical load is applied (Figure 3.13). This loading arrangement provides a horizontal line load on the top and bottom of the specimen. The loading creates very high compressive stresses in the specimen immediately under (and above) the line loads but theoretically uniform horizontal tensile stresses

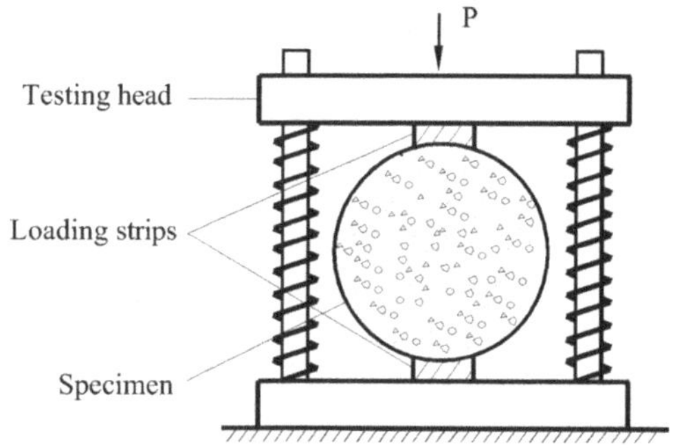

Figure 3.13 Indirect tension test (splitting test) with diametrically opposite line loads.

in the interior of the slice normal to the vertical plane between the line loads. The line loads are increased until the disc specimen splits open due to the internal tensile stresses.

The magnitude of the uniform tensile stresses, assuming linearly elastic, isotropic material behaviour, is:

$$\sigma = \frac{2P}{\pi Dh} = 0.636\frac{P}{Dh} \tag{3.5}$$

where:

P = the total line load (kN),
D = diameter of test specimen,
h = specimen (disc) thickness.

P is recorded when the specimen splits, and the corresponding indirect tensile strength (ITS) is computed from Eq. (3.5).

Experience has shown that the ITS determined this way is noticeably higher than the tensile strength of specimens of the same asphalt concrete mix determined by the direct tension test shown in Figure 3.8. This has to do with the effect of the compression region under (and above) the line loads, and that the location of the split section is predetermined by the location of the line load, while in the direct tension test, the weakest cross section along the tension specimen will fail. However, the difference has not been fully explained analytically. In practice, it has been suggested that the theoretical value obtained from Eq. (3.5) be reduced, somewhat arbitrarily, by multiplying the splitting test value with an empirical factor of magnitude (1-Poisson's ratio) to make the ITS value better agree with the tensile strength determined in the direct tension test.

Even though the tensile strength determined by the splitting test in combination with Eq. (3.5) may not give the correct tensile strength, the comparison of ITS values among specimens from various asphalt concrete mixes is useful and informative. For instance, the relative benefits of using various additives in the mix or the use of reinforcing fibres may be studied by using the simple splitting test. The test is also useful for studying the fracture surface and the adhesion (bonding) between aggregates and bitumen.

3.10 Additives to improve the hydro-mechanical behaviour of HAC

An asphalt concrete mix using straight (unmodified) bitumen as commercially produced may not satisfy the specific design requirements of a project. There are several additives used to improve the hydro-mechanical behaviour of HAC. Some refineries develop and produce modified bitumen to improve the flexibility, ductility, and cracking resistance of the asphalt concrete, for instance by modifying the straight B90 grade bitumen by adding styrene-butadiene-styrene (SBS) block polymer. Only a small amount is required to obtain significant effects, e.g., 4% of the bitumen weight. Nakamura et al. (2004) reported on the testing of a specially designed mix named "Superflex-phalt". By using additives in the bitumen, they reported that the HAC mix with the additive could tolerate tensile strains up to failure more than 30 times larger than when the mix with straight bitumen (B70) was used. Another method to improve the flexibility and ductility is to add natural or synthetic fibres to the mix. As discussed below, additives are frequently used to improve the bonding between bitumen and aggregates and the bonding at the interface with Portland cement concrete.

More than one kind of additive may be used to improve the hydro-mechanical behaviour. However, it should be noted that in addition to increasing the cost of the asphalt mix, additives may have negative effects on other hydro-mechanical properties of the mix than the ones improved. This must be investigated before the additives are used in the final mix for core construction.

It is not possible or advisable to give general specifications for what additives should be used to meet special requirements, as specific bitumen properties and available aggregate materials and project conditions vary from project to project. The decision to include additives to improve the hydro-mechanical behaviour of the asphalt mix should be based on documented test results. Test results on specimens prepared in the laboratory must always be validated by carrying out tests on specimens using the mix produced by the asphalt plant to be used for the project.

3.10.1 Additives to HAC to improve bitumen bonding to acidic aggregates

To obtain good aggregate-bitumen adhesion, alkaline aggregates like limestone or dolomite are usually preferred, as bitumen itself is acidic. When alkaline aggregates are mixed with acidic bitumen at high temperature, the bitumen develops physical adhesion to the surface of the aggregates, and chemical reactions take place. This may result in strong bonding between aggregates and bitumen.

For road and airfield pavements, good bonding (adhesion) between bitumen and acidic aggregate is essential to achieve satisfactory resistance against the effects of tyre tear and wear, inclement weather, and freezing cycles. Therefore, additives or anti-strip agents to improve bitumen-acidic aggregate bonding are required. The mechanisms and processes that create strong aggregate-bitumen adhesion are complex, and the bonding depends on various factors such as the lithology of the aggregate, the surface texture of the aggregate, the type of bitumen, and the asphalt concrete mix. Potential fracturing in the mastic between aggregates in asphalt concrete may be initiated by failure of the bitumen-aggregate adhesion, by low asphalt mastic cohesive strength or a combination, mainly depending on the load and environmental conditions. Presently, there are no generally recognized, quantitative methods to predict and evaluate the strength of bonding of various types of bitumen to aggregates of various types. Frequently, a dosage of Amin of 0.3%–0.5% of the bitumen weight is used, i.e., mainly based on experience not on comprehensive analyses.

For HAC used for asphalt facings of reservoirs, good bonding (adhesion) between bitumen and acidic aggregates is necessary to tolerate the required tensile strains without cracking and to obtain satisfactory resistance against the effects of inclement weather and ensure durability. A relatively high dosage (0.8% of bitumen weight) of the anti-strip agent was used to improve the bitumen-granitic aggregate bonding for the asphalt facing of the upper reservoir for the Yimeng pumped-storage project in China (2021). On the contrary, for HAC used in a dam core, the requirement of strong bitumen-aggregate bonding is not considered to be essential.

To study the effects of using additives in HAC to improve (strengthen) the adhesion to strip-prone acidic aggregates, Wang et al. (2010b) undertook a special research investigation using specimens produced with granitic aggregates that showed poor adhesion to bitumen. All the coarse aggregates and 50% of the fine aggregates were granitic materials. A typical aggregate gradation curve for asphalt concrete in a dam core was used in the experiments. The filler content was 14%, while the bitumen content (B70) was 6.5%. Unconfined compression tests at an axial strain rate of 1%/minute and beam bending tests at a strain rate of 0.6%/minute were performed at 15°C. One set of test specimens was prepared by adding the anti-strip agent Amin to the mix, while the other set did not have any addition. Both sets of compression and beam bending specimens were immersed in water for 15 days prior to testing and were exposed to 100 freeze-thaw cycles of –25°C to +25°C to simulate the effects of the harsh climate conditions at the dam site. Details of the study are presented by Wang et al. (2010b). The test results showed that there was no increase in the unconfined compressive strength when a dosage of 0.4% Amin was added to the bitumen, and no significant increase was observed in the bending strength and magnitude of bending strain to beam failure.

The reason for not obtaining any beneficial effects of the use of Amin may be that for HAC with air void content (air porosity) less than 3%, the permeability is extremely low, much lower than for road asphalt mixes. A thick bitumen coating is covering the aggregate particles in HAC, and water immersion and freeze-thaw cycles seem to have had no effect on the compression and bending strengths of the specimens even if strip-prone aggregates were used in the mix. The aggregates in HAC are enclosed in a thick film of bitumen preventing water from penetrating into the surface of the aggregate particles. Thus, the use of an anti-strip additive in the dam core, i.e., predominantly in compression and suffers little or no cracking, may give little or no benefit. However, for HAC used in dam facings that are subjected to both tension and bending and where fissures and cracks are more likely to occur, there may be a benefit of using an anti-strip additive.

For many ACED projects, alkaline aggregate materials may not be available locally, but only acidic rocks such as granites that are strip prone. A few examples are mentioned below.

For the Danghe ACED (58 m, 1974) in China, acidic natural Gobi gravels were used without applying an anti-strip agent but with a high bitumen content of 8.0% of the bitumen weight. The purpose of using the unusually high bitumen content was mainly to facilitate the asphalt layer compaction. The Danghe ACED has shown satisfactory performance since it was built in 1974.

The first ACED in China using an anti-strip agent (a dosage of 0.4%) was the Maanshan ACED (80 m, 2016) as the aggregates for the asphalt core were gabbro. The dosage of anti-strip agent was 0.3% for the asphalt core in the Nuer ACED (80 m, 2017) with granitic aggregates, but it was later increased to 0.5% due to the increase of acidic fines in the filler. The dosage of anti-strip agent was also 0.3% for the asphalt core with granitic aggregates in the Zungeru ACED (85 m, 2020), Nigeria. The aggregates for the asphalt core in the 58-m high Sidan ACED, Bali, Indonesia, are andesite, which is neutral, and the asphalt mix does not include an anti-strip additive. The dam was completed in September 2024.

Another example is the case of the granitic rock in the quarry for Chimney Hollow ACED in Colorado, USA. Initially, it was assumed that the bonding between the acidic aggregates and the bitumen would be poor and additives would be required. However, standard boiling water tests (ASTM) showed the bonding to be good, and no anti-strip additive was found to be required. The acidic, natural aggregates, especially the fine fraction of the aggregates, have good adhesion to bitumen, presumably because surface oxidation over a long time may have created a surface texture favourable for bitumen bonding. Thus, the degree of aggregate surface oxidation may be another variable that must be considered when evaluating bitumen-aggregate bonding.

Chapter 4

Effects of compaction method on the behaviour of HAC

The stress-strain behaviour of a laboratory-compacted specimen, and that of a specimen drilled out (extracted from) the compacted dam core may be very different, i.e., the case even if the asphalt concrete mix in the two specimens is the same and the air void contents (air porosity) after compaction are the same. This chapter presents and discusses the differences and provides experimental evidence and explanations for the differences in observed stress-strain behaviour. The differences must be considered and accounted for in the evaluation of stress-strain parameters for use in numerical stress-deformation analyses of the core, in observed vs. predicted dam performance, and in quality control during and after dam construction.

4.1 Various compaction equipment and procedures

Wang and Høeg (2002, 2009) investigated the effects of various laboratory compaction methods on the HAC stress-strain behaviour and compared the results obtained from laboratory-compacted and field roller-compacted specimens. All comparisons were based on results of stress-strain-strength behaviour of triaxial specimens in compression loading tests.

The asphalt concrete mix used for this study was the mix used in the construction of the Storglomvatn asphalt core embankment dam (case study in Section 10.3). The aggregates were crushed limestone, and the limestone filler content was 12% of the mineral weight. The bitumen grade was B180, and the bitumen content was 6.7% of total weight.

The asphalt concrete mix was prepared by following the procedure for standard Marshall mix design and testing as described in Section 3.4. The split steel mould used for preparing the triaxial specimens in the investigation was 100 mm in inner diameter and 250 mm in height. Four different laboratory compaction methods were used to prepare specimens to the same air void content (approx.1%). After compaction and cooling, the specimens were removed from the split mould and trimmed using a saw to a length of 200 mm. Details of the investigation may be found in the study by Wang and Høeg (2002, 2009).

4.1.1 Compaction of triaxial specimens by Marshall tamper in a 100-mm diameter mould

Application of the Marshall tamper (hammer) in a 100-mm diameter mould was one of the compaction methods to prepare the triaxial specimens used in the investigation. The asphalt mix was compacted in four consecutive layers, each originally approximately 60 mm thick. The top surface of each layer was compacted by 30 hammer blows (one blow per second). The diameter

DOI: 10.1201/9781315177786-4

of the hammer is 98.4 mm, which is only 1.6 mm smaller than the inner diameter of the compaction mould for the triaxial specimens.

4.1.2 Compaction by the Gyratory method

The 250-mm high cylindrical mould was filled in one layer. The mix was then compacted by a combination of sustained axial stress and shear stress as the mould was gyrated in a circular motion maintaining a small 2°–3° angle with the vertical axis. Shear movements under sustained axial stress is a very efficient compaction method for granular materials, and the compaction was stopped when the mixture reached the specified density that corresponded to an air void content (air porosity) of 1%. For the specimens prepared in this study, the axial stress used was 0.4 MPa and the gyration frequency 30 rpm.

4.1.3 Static compaction method

The mix was placed in the 250-mm high cylindrical mould in four consecutive layers, each approximately 60 mm thick. A bullet-nosed steel rod 10 mm in diameter was used to penetrate each layer 25 times ("rodding") to level the layer and to get air out of the mix. When the four layers had been placed, a rigid steel disc with a diameter of 98.4 mm was placed on top and bottom of the specimen. Then, an axial force, corresponding to an average axial stress of 15 MPa, was applied for 2 minutes.

4.1.4 Vibration method

The mix was placed and compacted in four consecutive layers. Each layer was placed approximately 60 mm thick and compacted for 30 seconds by a vibrator with a frequency of 50 Hz, with a weight of 70 kg, and a flat circular loading plate with a diameter of 98.6 mm.

4.1.5 Field roller compaction

The field specimens (diameter of 100 mm) were carefully drilled out of a test section compacted by field rollers (Saxegaard, 2002). The 15-m long test section simulated the conditions for a 0.5-m wide asphalt concrete core wall in an embankment dam with a 1.5-m wide gravel transition zone (Zone 2) on each side of the core. The asphalt concrete was placed in one layer approximately 250 mm thick which after compaction reduced to approximately 200 mm. Compaction of the asphalt was achieved by a 0.8-tonne vibratory roller moving along the axis of the dam at a speed of approximately 2 m/minute, using three passes with vibration and two passes without. This is the same compaction procedure and equipment as used for the core in ACED field construction (see Chapter 6).

Three (3) specimens were prepared and tested for each of the four different laboratory compaction methods and for the field roller-compaction method. All triaxial specimens had an air void content (air porosity) of approximately 1% after compaction.

4.2 Effects of compaction method on triaxial stress-strain behaviour

The triaxial test procedure was the same as that described in Section 3.7. The test temperature was 5°C. The constant axial strain rate used was 2%/hour (displacement rate 0.067 mm/minute) keeping the radial confining stress constant equal to 1.0 MPa. The average triaxial stress-strain

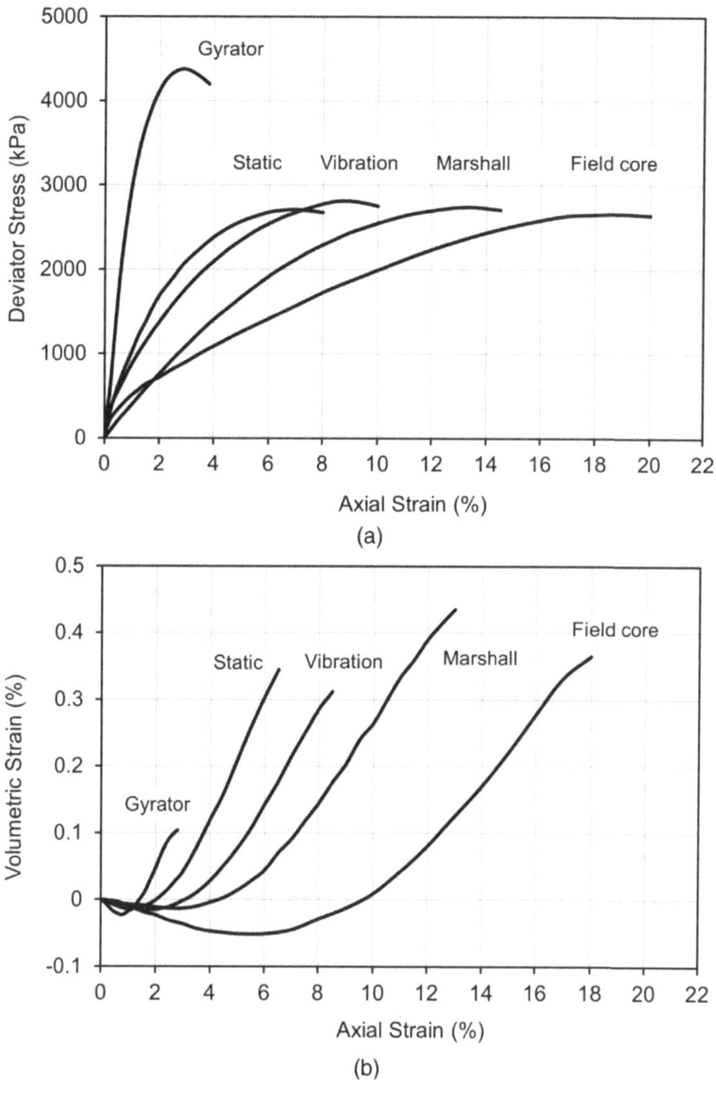

Figure 4.1 Average curves for (a) deviator stress vs. axial strain, and (b) volumetric strain vs. axial strain for the five different methods of compaction. All specimens were initially compacted to an air void content (air porosity) close to 1.0%.

curves for each of the five different compaction methods are shown in Figure 4.1, which presents the deviator stress and volumetric strain vs. the axial strain.

The results show that the asphalt specimens prepared by the five different compaction procedures to the same initial air void content exhibited very different stress-strain behaviour. The gyratory laboratory compaction method gave specimens of much higher stiffness and strength than the other methods. The differences in strength (max deviator stress, i.e., $\sigma_1 - \sigma_3$) among the specimens compacted by the other laboratory compaction methods and the field specimens were relatively small, but the differences in compression modulus and axial strain to failure were large. The initial secant modulus up to 1.0% axial strain varied from about 100 MPa for the specimens compacted by static laboratory compaction to 40 MPa for the specimens compacted by Marshall hammer laboratory compaction. The corresponding average axial strain at failure for the two methods was 7% and 14%, respectively. The gyratory laboratory specimen gave an average secant modulus of about 300 MPa and axial strain at failure of 3%. The initial secant modulus for the gyrated specimen is six times that of the field roller-compacted specimen, and the axial strain at failure is only 1/6 (3% vs.19%).

The aggregate arrangement and structure after compaction and shearing were studied by cutting (sawing) through the specimens along the potential failure planes after the triaxial tests were completed. The sections clearly showed a higher degree of aggregate interaction and interlocking for the specimens compacted by the static method than those compacted by the Marshall hammer or the vibration method. The weakest interlock was exhibited by the specimens compacted by the Marshall hammer. This observation agrees with the finding that the average initial secant modulus was 102, 88, and about 40 MPa, respectively, for the groups of specimens compacted by the three different laboratory methods. At large axial and shear strains, the initial aggregate structure is no longer intact and does not have much influence on the peak strength, which was found to be rather similar for the three groups of specimens.

The differences in aggregate structure, grain skeleton arrangement, and degree of interlocking achieved during the various methods of compaction explain the large differences in observed stress-strain behaviour. The elongated and plate-shaped aggregates have significant effects on the structure of the aggregate skeleton, which in turn strongly depends on the method of compaction. This is discussed in Section 4.5.

The gyrator exerts a vertical stress as well as a shear stress on each end of the specimen during the compaction process. There is a counter clockwise torsional moment on the rigid plate at one end and a clockwise moment on the other during the gyration around the vertical axis. The elongated aggregates in the mix tended to be arranged in the radial and tangential directions with the specimen axis as a reference. The resulting aggregate interlock obtained with the gyrator compaction method is much stronger and stiffer than in the other laboratory-compacted specimens and the roller-compacted field specimen.

For the Marshall hammer, vibration, and static compaction methods in the cylindrical mould (diameter of 100 mm), there is a rigid circular plate on the specimen top surface covering almost the full circular cross section of the compaction mould. There is no lateral movement of mass during compaction due to the rigid steel walls and very little surface heave because the top plate covers almost the complete surface of the specimen. There is no kneading action and no rotation of stresses. The compaction of triaxial specimens in the laboratory does not at all simulate the movements and behaviour under the roller in the field.

Additional triaxial tests were run on specimens prepared by the static compaction method to investigate the influence of the duration of applied axial compaction stress on the resulting stress-strain behaviour. The axial compaction stress was kept constant at 10 MPa for all the specimens but the durations were 2, 4, 6, and 10 minutes respectively, for the various specimens. The resulting compression modulus increased from 27 to 54, 115, 151 MPa, and the strength increased from 2.0 to 2.5, 2.8, 3.0 MPa when the duration of the static compaction stress increased. The axial strain at failure decreased from 13% to 9.5%, 7.0%, 5.0% as the duration increased. The tests clearly showed that with increased duration of static compaction stress, there is more time for viscous flow, aggregate rearrangement, interlocking, and increase in specimen stiffness and strength to occur.

4.3 Limitations of Marshall compaction method for triaxial specimens

The test results presented in Section 4.2 indicate that the stress-strain behaviour of triaxial specimens compacted in the cylindrical mould with an inner diameter of 100 mm and using the Marshall hammer may be close to that of asphalt specimens drilled out of the core in the field. However, further experimental studies have shown that depending on the characteristics of the aggregates and the asphalt concrete mix, the differences can be much larger. Triaxial tests performed on the asphalt mix for the Greater Ceres ACED, South Africa (Jones et al., 1999) and on the asphalt mix for the Maopingxi ACED, China (Hu et al., 2001) showed very significant differences between the stress-strain behaviour of field core specimens and the Marshall laboratory-compacted specimens although they initially had virtually the same air void content. The tests showed that the laboratory prepared specimens gave an initial secant modulus up to 1% axial strain about three times higher than those of the specimens compacted by rollers in the field.

For the Dongtang ACED, China, the initial modulus of the triaxial specimens compacted by the Marshall hammer in the cylindrical mould was found to be about four times higher than that of the specimens drilled out of the asphalt core, while for the Yatang Dam, the ratio between the initial modulus values was only two. It was concluded that it was the significant amount of elongated and flaky particles (flakiness) that caused the large differences in stress-strain-strength behaviour between the Dongtang ACED laboratory-compacted and field-compacted specimens. The ability of the triaxial specimen compaction method to produce specimens that exhibit field specimen stress-strain behaviour depends on the asphalt mix proportions and the aggregate characteristics. Further studies were performed to obtain in the laboratory specimens that exhibited the same stress-strain behaviour as specimens compacted by field rollers.

4.4 Laboratory method to better simulate the results of field compaction

Experience has shown that for the asphalt mix to be used in the dam core to have an air void content less than 3% after field roller compaction, the air void content of the standard Marshall specimen after standard laboratory compaction should be less than 2%.

The Marshall hammer compaction in the 100-mm diameter mould for triaxial specimens did not give specimens with the stress-strain behaviour like the behaviour of field-compacted specimens for an asphalt mix like the Dongtang ACED mix. In the case of Aikou ACED (Wang, 2004), the differences between the measured moduli on laboratory – and field – compacted specimens were even larger, and the reason for that was studied further. The aggregate source for the asphalt core in Aikou ACED is sound limestone, and triaxial specimens with 6.5% bitumen content were prepared. It was proved difficult to get an air void content in a standard Marshall specimen as low as 2% even when the number of hammer blows was increased. The Marshall hammer compaction with 30 blows gave an air void content of the triaxial specimen of about 4.0%. To explain this unusual behaviour, several Marshall specimens of the Aikou mix were tested in triaxial compression. The axial displacement rate was 0.2 mm/minute (i.e., axial strain rate of 0.1%/minute) at 16°C, and various magnitudes of confining stress were applied. The stress-strain curves recorded are shown in Figure 4.2.

The stress-strain results showed very high initial compression modulus values and very small axial strains at failure. To achieve 2% air porosity, many more blows would have to be used leading to stiff and brittle specimens, not well suited for use in a hydraulic barrier that

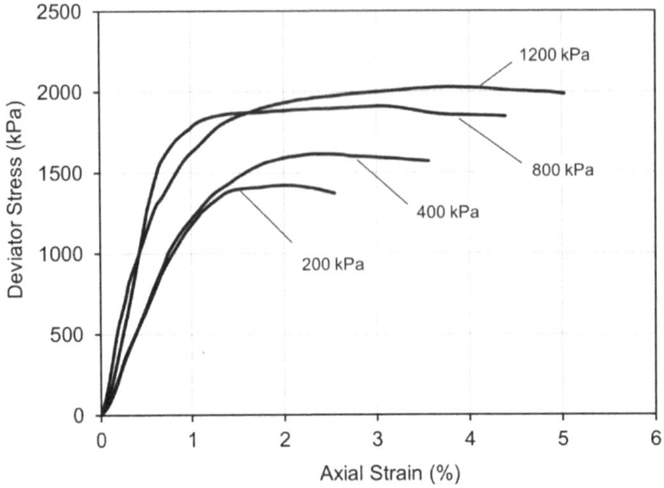

Figure 4.2 Deviator stress vs. axial strain for triaxial specimens compacted by the Marshall hammer in the 100-mm diameter mould for the Aikou ACED using various values of confining stress.

should be flexible and ductile. The findings at the time caused uncertainty about the suitability of the selected aggregates and the stress-strain curves for the asphalt core of the Aikou ACED (Figure 4.2). The uncertainties triggered further studies.

A new laboratory compaction mould and method to better simulate the conditions for asphalt concrete compacted by rollers in the field was designed for asphalt mixes like those for the Dongtang and Aikou ACEDs. An important parameter in the study was the effect of elongated and flaky aggregate particles, i.e., the flakiness index (FI), on the stress-strain behaviour.

4.5 New laboratory compaction mould and procedure

Based on the observations and experience with triaxial specimens prepared and compacted in the cylindrical 100-mm diameter Marshall mould, the second author designed a 120-mm wide, 250-mm long, and 250-mm high rectangular mould as shown in Figure 4.3. A 10-mm thick rubber lining was glued inside the two long walls to allow some lateral movement of the mix inside the mould during compaction. Holes with a diameter of 5 mm were evenly distributed, 200 mm apart, on all four wall areas, to allow air to escape during the compaction process. A small steel spring was placed between the two nuts on each screw at the top and bottom that connect the longitudinal walls to the short end walls to allow some lateral flexibility of the mould. The flat circular tamping surface of the standard Marshall hammer was modified to a rectangular shape with a dimension of $70 \times 110\,mm^2$. The equipment and compaction procedure allow a small surface heave outside the tamping surface during compaction as the hammer is moved along the length of the mould in small steps.

The mould sits on a 10-mm thick rubber slab, and the hot asphalt concrete mix is compacted in four consecutive layers, each originally about 60 mm thick. The hammer is moved along the length of the mould in steps of ca. 35 mm, which is half of the hammer width, with one drop of hammer per step. For each new layer, 15 passes with the compaction hammer are used. Several

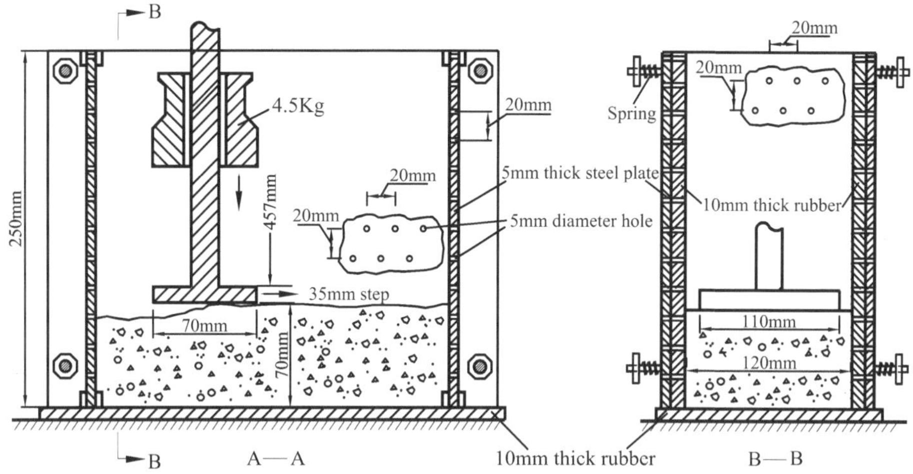

Figure 4.3 Equipment developed to better simulate field compaction (SFC) of asphalt core.

trials were run with this new type of compaction mould to better simulate field compaction (SFC) conditions. This eliminates some of the shortcomings of the current compaction equipment and procedure. The first tests were run on the Aikou dam asphalt core mix that had given surprising and unsatisfactory results when the conventional compaction equipment and procedures were used (100-mm inner diameter mould and 98.5-mm diameter circular tamper).

After compaction and cooling, two test specimens (diameter of 100 mm and about 250 mm in length) were drilled out of the rectangular asphalt block in the compaction mould and trimmed using a saw to a length of 200 mm. The air porosity of the triaxial Aikou specimens prepared in the SFC mould was a little less than 2.0%. As described above, when the Aikou mix was compacted by the Marshall hammer in the conventional cylindrical mould for triaxial specimens (denoted by M), it was difficult to compact to a void content less than 4%. The comparisons of the triaxial stress-strain curves for the M and SFC specimens, respectively, are shown in Figure 4.4.

From Figure 4.4, one may observe that the stress-strain curves are very different for the M and SFC specimens. The SFC curves are much more like the curves from the specimens compacted by the field roller than the results from the M specimens. The average initial secant modulus of the M specimens is about three times that of the SFC specimens, and the average axial strain at failure is about ¼ that of the SFC specimens.

For an asphalt mix with a significant amount of elongated and flaky aggregate particles (high FI), the M compaction equipment and method cannot give triaxial specimens with air void content less than 2% and stress-strain behaviour like that of specimens roller-compacted in the field. The SFC compaction equipment and method gives laboratory specimens with stress-strain behaviour much closer than the M specimens to that of the asphalt compacted by rollers in the field. This has been demonstrated in many projects by comparing the stress-strain behaviour of SFC specimens, M specimens, and specimens drilled out of the roller-compacted dam cores. Asphalt concrete core mixes for, e.g., Dongtang, Aikou, Xieka, and Guangmaozhou ACEDs in China have been investigated. Based on these studies, the SFC compaction equipment and method became one of the specified compaction methods for HAC in the Chinese testing code (NEA, 2019).

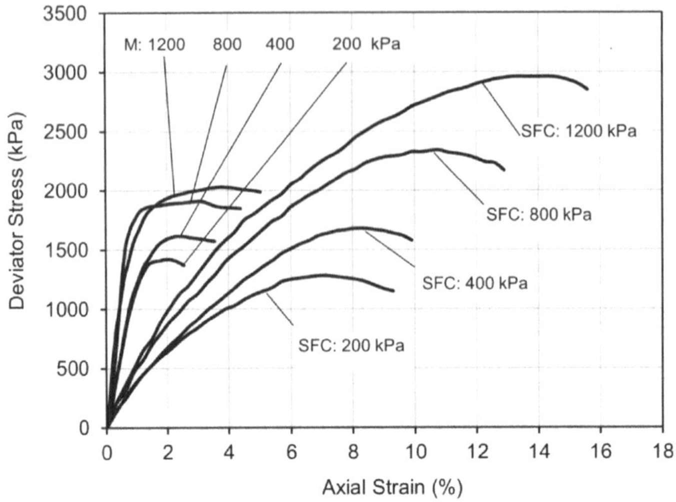

Figure 4.4 Comparisons of stress-strain curves for triaxial specimens compacted by Marshall hammer in a 100-mm diameter mould (M) and by the simulating field compaction (SFC) mould and method. Tests were performed on the asphalt mix used for the Aikou ACED asphalt core.

4.6 Recommended laboratory compaction for HAC triaxial specimens

The method of laboratory specimen compaction has very significant effects on the stress -strain behaviour obtained in triaxial compression tests. This must be considered when stress-strain-strength parameters for use in numerical (e.g., finite element) stress-deformation analyses are selected to predict field performance during design or, to compare with observed field behaviour during quality control. When the crushed aggregates (larger than 6.3 mm) include many elongated, flaky particles (i.e., has high FI), the conventional Marshall compaction method in a 100-mm diameter mould should not be used. In such cases, the SFC method should be applied with the modified laboratory compaction mould and compaction procedure.

The first generation of SFC compaction moulds and procedures has been developed and has been found to be appropriate and very useful. With more experience, the equipment and procedures may be further improved to simulate field conditions even better. The SFC compaction procedures are more elaborate and time consuming than the traditional Marshall compaction procedures but yield much more representative specimens to model field stress-strain behaviour.

The authors make the following recommendations and criteria related to aggregates and suitability testing for HAC mixes:

- The upper limit for the FI for the coarse aggregates larger than 6.3 mm to be used in a HAC mix is 35.
- For FI lower than 20, the traditional Marshall compaction equipment and procedure may be used, but the SFC equipment and procedure gives more realistic stress-strain results that simulate the field behaviour.

- For $20 < FI < 35$, the traditional Marshall equipment and procedure should not be used. The SFC equipment (or similar) and procedure should be used to get an air void content (air porosity) lower than 2% and realistic stress-strain behaviour simulating the field behaviour.
- $FI > 35$ is not acceptable for use in the HAC mix.

To reduce the FI value for the aggregates larger than 6.3 mm, the project crusher must be modified or replaced or a new source for rock aggregates that give lower FI must be used.

Chapter 5

Cracking resistance, stress relaxation, and self-healing

When designed and constructed according to the technical specifications with an air void content of not more than 3%, the core is virtually watertight with a permeability coefficient of about 10^{-10} cm/s. The core should be designed to remain watertight without detrimental crack formation during construction, impoundment, and dam operation.

5.1 Changes in specimen permeability during triaxial compression tests

Høeg (1993) reported the results of permeability tests on triaxial specimens of HAC that were subjected to axial compression. After loading, small specimens were cut out in the region of high shear strains along the inclined potential failure planes in the triaxial specimens and were tested in a permeameter (Figure 3.3). Although the volumetric strains during shear dilation had shown very little expansion of the triaxial specimens during the compression tests, the permeability coefficient of the small specimens showed an increase by a factor of 100 or more compared to the permeability of the unloaded triaxial specimens. This large increase was explained by the existence of a set of communicating (interconnected) fissures or small cracks that had developed in the specimens during the triaxial compression tests. The dramatic increase in permeability only occurred in specimens brought close to shear failure (shear strength) for the specimens. No significant increase in permeability was detected until about 90% of the shear strength of the triaxial specimen was mobilized.

Qu et al. (2004) studied the issue of the increased permeability in more detail. They performed permeability tests on 65-mm high specimens cut from the middle part of 100-mm diameter and 200-mm long triaxial specimens that first had been subjected to stress levels corresponding to 25%, 50%, 75%, 100%, 75% (post-peak), and 50% (post-peak) of the peak strength. The bitumen content (B70) in the specimens was 7.8%, and the tests were run at 7°C without applying radial confining stress (i.e., unconfined compression tests). The test results agreed with the earlier findings and showed that the 65-mm high specimens were practically watertight up to a stress level close to the shear strength of the asphalt concrete. Zhang et al. (2013) reviewed the work by Qu et al. (2004) and performed permeability tests on horizontal slices (discs) of 20-mm thickness cut from the middle part of triaxial specimens that first had been subjected to shear stresses up to 80%, 90%, and almost 100% of the shear strength. The bitumen (B90) content in the specimens was 6.5%. The triaxial compression tests were run at an axial strain rate of 0.1%/minute with a constant radial confining stress of 1.0 MPa at 15°C. Permeability tests were carried out on the disc specimens under a water pressure of 1.0 MPa with the direction of flow along the axis of the triaxial specimens from which the discs had been cut. The permeability

DOI: 10.1201/9781315177786-5

coefficient for discs cut from the unloaded triaxial specimen was found to be approximately 5×10^{-10} cm/s, and the permeability decreased with time during the permeability testing. The corresponding permeability coefficients, determined from discs cut from the specimens subjected to about 90% of the peak strength, were found to be approximately 10–15 times higher presumably because communicating fissures had developed in the loaded triaxial specimens.

In a separate study, Ning et al. (2024) performed a triaxial compression test on an asphalt specimen drilled out of the core of the Gushan ACED, China, during construction. They simultaneously monitored the void changing process in the specimen and any fissure/crack development by using computer tomography (CT) during the triaxial testing. The CT revealed that the voids underwent a volume reduction and expansion process that generally corresponded to the volumetric strain reduction and expansion process recorded by the sensors used in the triaxial test measurements. Just before the peak deviator stress was reached, the voids and fissures started to interconnect. After the peak was reached, the interconnection of voids and the development of cracks grew, and the deviator stress decreased as the shear resistance decreased (strain-weakening).

Since 2019, researchers at Xi'an University of Technology have conducted a series of triaxial compression tests applying various confining stress levels on laboratory – prepared specimens. The bitumen content (B110) in the specimens was 7.0%, and the triaxial tests were run at an axial strain rate of 0.1%/minute at 10°C. The confining stress levels applied were 0.5, 1.0, 1.5, and 2.0 MPa, respectively. Triaxial specimens (diameter of 100 mm and 200 mm in height) were first run up to various axial strain levels and then dismounted. From the middle part of each specimen, four horizontal slices (discs) of 30 mm thickness were cut out to determine the permeability after the triaxial specimens first had been subjected to large axial strains. The recorded vertical compressive strain when the permeability for the various confining stress levels increased significantly (by more than an order of magnitude) is shown in Figure 5.1. The results appear very consistent and show the effect of the magnitude of radial confining stress on the magnitude of axial strain when the permeability clearly increased.

The tests showed that even if there was no significant increase in volumetric strain measured during triaxial compression, the asphalt concrete increased its permeability significantly

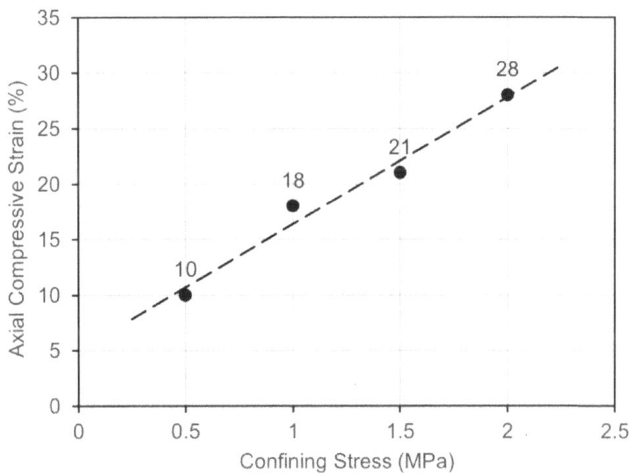

Figure 5.1 Axial compressive strain when the permeability coefficient increased significantly vs. confining stress level. The bitumen content (B110) was 7.0% in these tests.

with increasing shear strains. Shear and tensile strains are not uniformly distributed inside the specimens, as the largest strains occur between the largest particles inside the specimen. The large shear strains in these locations cause the asphalt mastic to peel off the large aggregates, and fissures and cracks initiate, propagate, and interact to significantly increase the permeability (Ning et al., 2024).

5.2 Cracking and self-healing behaviour

For a dam located in a seismic region, the slender asphalt core must be able to follow the cyclic and permanent strains imposed by the dam embankment on the core without undergoing excessive cracking and material degradation. If the computed results in the design analyses indicate that the core may undergo excessive deformations and cracking during strong earthquakes, special design measures must be implemented to mitigate the damage and increase the dam's earthquake resistance (Chapter 9). Asphalt concrete exhibits crack self-healing (self-sealing) and stress relieving properties due to the viscous material behaviour, and the process of self-healing has been studied for several different HAC mixes at various temperatures and stress levels.

A self-healing test was designed to simulate the situation in an asphalt core. A specimen of cross section of $70\times70\,mm^2$ and 100 mm in height was used. An almost horizontal crack was created at mid-height of the specimen by the splitting method using a 10-mm diameter steel bar as strip load at a temperature of –5°C (see Section 3.9). Then, the cracked specimen was mounted in the test apparatus as is shown in Figure 5.2, and a force was applied vertically corresponding to the axial compressive stress in the dam core. The water pressure difference between the upstream and downstream sides of the specimen was 1.0 MPa in the tests reported in this chapter. Typical leakage vs. time relationships is shown in Figure 5.3 for three different vertical stress levels.

For an asphalt core under vertical compression stresses, Figure 5.3 shows that crack self-sealing occurred rather quickly with time. Within only 5 hours, the rate of leakage dropped by

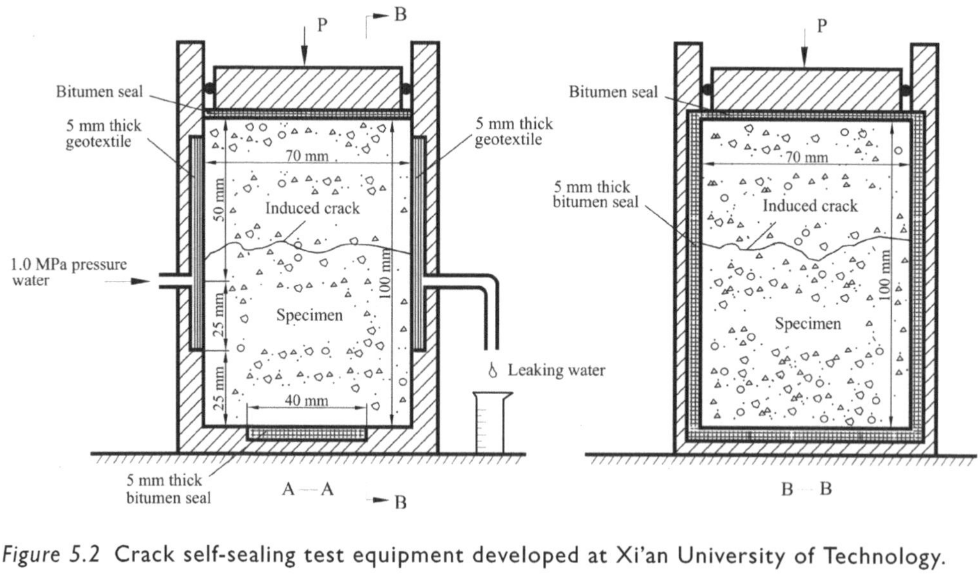

Figure 5.2 Crack self-sealing test equipment developed at Xi'an University of Technology.

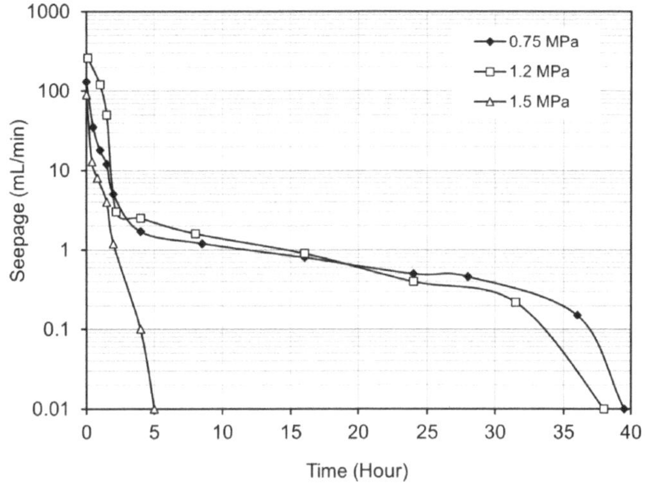

Figure 5.3 Self-sealing tests. Leakage vs. time under vertical stresses of 0.75, 1.20, and 1.50 MPa, respectively.

one (1) to four (4) orders of magnitude depending on the imposed compressive stress level. The leakage rate continued to decrease 1–2 orders of magnitude over the following 20–40 hours, but at a slower rate.

Similar self-healing tests were performed on the asphalt core mix for the Yele ACED, China (see Section 10.4 and Zhang et al., 2013). The bitumen content (B70) was 7.0%. One specimen was tested with 1.0 MPa vertical stress at 15°C, and the other with 0.5 MPa at 25°C. The leakage rates for the two specimens decreased to about the same value of 0.05 mL/minute after 16 hours of self-sealing.

Special investigations were performed to determine the tensile strength of the specimens after a crack had formed and the crack had been virtually sealed. After the self-healing tests had been performed with a vertical stress of 1.0 MPa for 24 hours at 7°C, the regained tensile strengths of the Yele ACED specimens were determined by performing splitting tests. The strengths of intact (no cracks) specimens were also tested by splitting, and a comparison of tensile strength results could be made. The average regained strength for three groups of three specimens each was 55%. These findings agree with the earlier results presented by Saxegaard (2003) who used a bitumen of grade B180. In his studies, the imposed vertical stress was 0.26 MPa on cubical specimens of 75 mm on the side, and the testing temperature was 5°C and 20°C, respectively. In only 4 hours, more than 50% of the original tensile strength was regained at 5°C and 75% at 20°C. Thereafter, the regaining of strength continued but at a slower rate.

The unique self-healing ability of asphalt concrete may eliminate the need for remedial measures due to possible crack formation caused by temporary flooding or earthquake shaking, as the asphalt core is in sustained compression and will self-heal with time.

5.3 Ageing, weathering, and durability of dam core

The durability of the asphalt core is important as ACEDs can be designed for a service life of 150–200 years. The service life of an asphalt road pavement may typically be less than 10–15 years, The service life for asphalt facings for canals and dams is 30–40 years. The loading, ageing, and weathering situation for a dam core is much less severe than for a pavement, and the service life is therefore much longer.

The service life of the HAC barrier in a dam is related to the "ageing" of bitumen during the hot asphalt concrete production and the dam barrier construction. The HAC mixes are usually kept in the temperature range of 150°C–170°C during the production process and that causes a primary "ageing" (hardening) due to the oxidation and volatilization of the bitumen. As stated in Table 2.2, the drop in penetration value for bitumen B70, B90, and B110 was 39%, 43%, and 45%, respectively, after heating the bitumen at 163°C for 5 hours to simulate the ageing of the bitumen during HAC production and dam core construction. Kawamura (1989) investigated the ageing of bitumen and showed that the drop in penetration after hot production was found to be in the range 15%–40%. Wang et al. (2013) investigated the ageing during construction of the B70 grade bitumen used for two HAC projects in China. The reduction in penetration value was 32% for bitumen extracted from the asphalt concrete after the hot production process. Ideally, to reduce loss in the penetration value of bitumen during the construction period and lengthen the service life of HAC, techniques should be developed to produce asphalt concrete at much lower temperature but at the same time maintain the unique hydro-mechanical and ductility properties of the HAC now produced at high temperatures.

The design life of asphalt concrete is strongly dependent on the weather and climatic conditions during construction and operation. For road asphalt pavements, the basic principle is to design a strong aggregate structure where the bitumen is bonding together the aggregates in the asphalt concrete. To achieve this, the road asphalt concrete is designed with lower fine aggregate content, lower filler content, lower bitumen content, and higher permeability than used for HAC.

The required air void content (air porosity) for road asphalt is more than 3.0% because porous road asphalt pavements are exposed to varying weather conditions and contain open and interconnected air voids that allow water to permeate the asphalt concrete material. Road pavements are subjected to continuous dynamic loading by vehicle tyres. The water inside the asphalt pavement, especially under dynamic loading, has a stripping action on the bitumen bonding between aggregates. Under such circumstances, road asphalt pavements are expected to have a service life of maximum 15 years.

For asphalt facings of embankment dams, the impervious layer of the facing is normally designed with a thickness between 5 and 10 cm (ICOLD, 1999). The asphalt facing can be divided into three zones: the exposed zone, which is near the top of the dam where the reservoir water rarely reaches; the middle zone of the available depth of the reservoir; and the underwater zone. An investigation was made to study the ageing that had occurred in the three zones of an asphalt facing used as water barrier for a daily pumped-storage reservoir after having performed for 13 years (Wang et al., 2013). The most severe ageing occurred in the most exposed zone but was found to be only "skin deep" (0–10 mm) in the asphalt facing. The properties of the asphalt facing under the "skin" had been only insignificantly affected after 13 years of service. Some of the exposed facings showed that only the outer layer of the facings in the exposed zones may need to be rehabilitated after 40 years (ICOLD, 1999; Schönian, 1999; Lengfeld and Müller, 2018).

An asphalt core that usually is in the centre region of the dam embankment is ideally protected and is in an almost constant temperature environment, only subjected to slow seasonal reservoir cyclic loading. If the asphalt core is designed and constructed according to technical specifications, the service life could be hundreds of years.

5.4 HAC stress relaxation behaviour

Asphalt concrete possesses viscous relaxation properties that can relieve imposed tensile stresses and stress concentrations that may have developed during construction and operation or have been caused by earthquake loading or foundation movements. The relaxation behaviour has been investigated using the equipment shown in Figure 3.8. Special stress relaxation tests were carried out at temperatures of 10°C, 5°C, and 0°C, considering the low temperatures in the top part of the asphalt core for a dam in the sub-arctic region. After the specimen had been kept at a specified temperature for at least 1 hour, it was quickly stretched to a specified length (specified tensile strain) in less than 3 seconds, and then, the strain was kept constant in the direct tension apparatus (Figure 3.8). The tensile stress required to maintain the specific strain level gradually decreased with time, and the relaxation modulus is defined and expressed as:

$$E(t) = \sigma(t)/\varepsilon_0 \tag{5.1}$$

where $E(t)$ = relaxation modulus; $\sigma(t)$ = relaxation stress level; ε_0 = strain kept constant.

$$E(t)\Big|_{t\to 0} = \sigma(t)/\varepsilon_0\Big|_{t\to 0} = E_0 = \sigma_0/\varepsilon_0 \tag{5.2}$$

where σ_0 and E_0 are defined as initial stress and initial modulus, respectively.

Typical relaxation modulus values plotted vs. the logarithm of time are shown in Figure 5.4 for HAC.

Figure 5.4 shows that the tensile stress and relaxation modulus are strongly affected by temperature and time. For instance, the initial modulus of more than 2,000 MPa at 5°C decreased to lower than 200 MPa in less than 10 minutes.

A simple field test demonstrated the flexibility, ductility, and viscous relaxation properties of HAC (Figure 5.5). A beam made of HAC was lifted and bent by a shovel as shown in the photo. The beam kept its bent, plastically deformed shape without cracking.

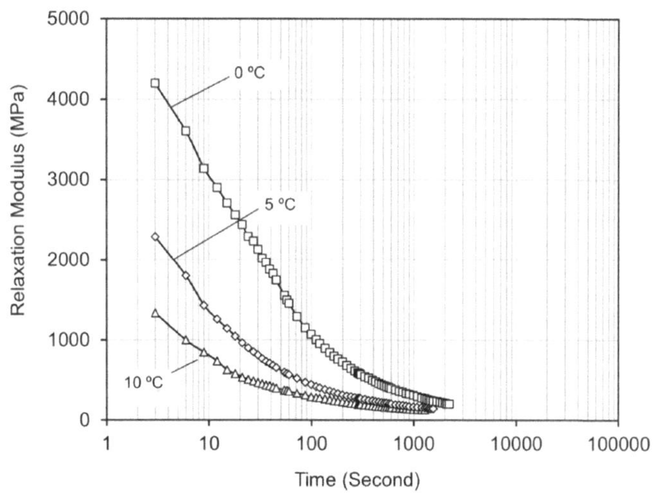

Figure 5.4 Typical relaxation modulus vs. time at various temperatures for hydraulic asphalt concrete (HAC) (Zhang et al., 2013).

Figure 5.5 Demonstration of flexibility and ductility without cracking of an asphalt concrete beam in a field test (ICOLD, 2018).

HAC is a "forgiving" material, relieving itself of stress concentrations and tensile stresses without cracking. It is well suited as a material to be used in watertight barriers in embankment dams.

5.5 Cracking caused by rapid cooling to low temperature?

Under severe climate conditions, the top part of the asphalt core may be subjected to freezing temperatures, and a concern may be raised that a rapid drop in temperature can create thermal stresses that could cause cracks to develop. For the design of asphalt concrete facing or canal lining, this situation is of more concern and has been studied.

A special low-temperature crack-opening test is used to evaluate the resistance to cracking using the equipment shown in Figure 3.8. During the test, the length of an asphalt specimen is kept constant, while the temperature is decreased until cracking occurs. The starting temperature in the test is usually 10°C, and the rate of temperature decrease is 30°C per hour. A typical response curve for the tensile stress developed in a constrained specimen is shown in Figure 5.6. When the temperature was reduced from 10°C to –7°C, there was no tensile stress built up in the specimen. This implies that the potential tensile stress caused by the temperature decrease was dissipated by the high stress relaxation rate in the asphalt concrete. When the temperature was reduced from –7°C to –21°C, the tensile stress in the specimen started to build up and the rate increased with the temperatures below –7°C. When the temperature was reduced below –21°C, the tensile stress increased linearly with the temperature reduction until a crack opened. The transition point on the curve, in this case at –21°C, is defined as the stress relaxation limit for this asphalt concrete. The stress and temperature at crack-opening (4.2 MPa and –42.7°C in the example in Figure 5.6) and the stress relaxation limit are important parameters used to evaluate the behaviour of an asphalt canal lining or upstream dam facing at low temperatures.

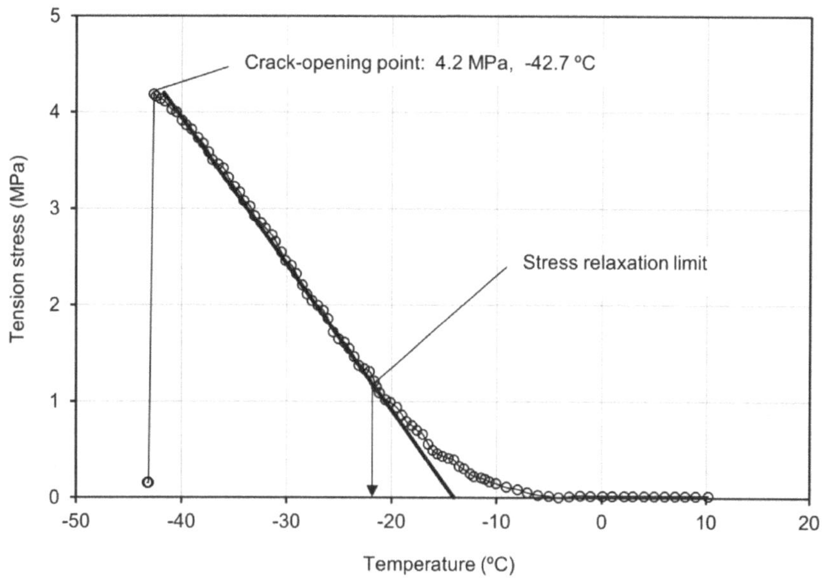

Figure 5.6 Typical crack-opening test for hydraulic asphalt concrete (HAC) (Zhang et al., 2013).

The top part of an asphalt core is normally protected by a few metres of fill or maybe by the base of a parapet wall. The temperature in the core may therefore be higher than −10°C, and the temperature reduction process may be slower than that in the test shown in Figure 5.6. Furthermore, the top of the asphalt core can shrink freely in the vertical and upstream-downstream directions and is only restricted in the cross-valley direction. The test results in this example show that there were no critical stresses or cracking produced in the core due to the viscous relaxation properties, and there is no concern about low-temperature cracking of the asphalt core. However, low-temperature cracking needs to be investigated for asphalt road pavements or asphalt facings when the projects are in very low-temperature regions (Wang, 2008).

5.6 No effects of acidic water on hydro-mechanical properties of HAC

Research has been done to study possible effects of acid rain on the properties of asphalt concrete used for roads and airport pavements. Insignificant effects have been found by these investigations. However, no research has so far been done to study the effects of strongly acidic water on the hydro-mechanical properties of HAC. For the storage of mine tailings, asphalt concrete is being used in dams with impervious upstream facings or central cores, where the asphalt concrete is exposed to very acidic mine drainage (AMD). The uncertainty is whether the properties of HAC deteriorate when exposed to strongly acidic water.

5.6.1 Test program with HAC in acidic water

Feng et al. (2020b) therefore undertook an extensive experimental study of this issue. HAC specimens with air void contents (air porosity) of less than 2% were immersed in acidic water with pH = 2, 4, and 6, for up to 66 months (5.5 years), and indirect tensile (splitting) tests were carried out on specimens at various times during the immersion period.

Standard Marshall asphalt specimens were prepared (see Section 3.4). The aggregates were limestone, and the bitumen content (B90) was 6.5%. The asphalt specimens were compacted with 35 blows of the Marshall hammer on the top and bottom faces to yield specimens with porosity inside the range 1%–2%. Indirect tensile tests were carried out at a displacement rate of 1 mm/minute at 20°C to measure the tensile (splitting) strength and the corresponding tensile strain up to when the crack opened.

The specimens were separately put in polyethylene (PE) plastic containers filled with acidic water with various pH values. The acidity of the water was kept constant at 20°C. At predetermined times (every 3 months during the first 2 years), asphalt specimens were taken out of the containers for indirect tensile strength testing. Each specimen was only subjected to the indirect tensile test once, and there was no effect on the recorded splitting strength of prior testing of the same specimen.

Neutral water (pH = 7) had insignificant effect on the asphalt tensile strength with time, and the test results for the specimens in the neutral water were taken as reference values for the specimens immersed in the acidic water with pH = 2, 4, and 6, respectively. Figure 5.7 shows the tensile strength ratios with time of asphalt specimens immersed in acidic water. The strength

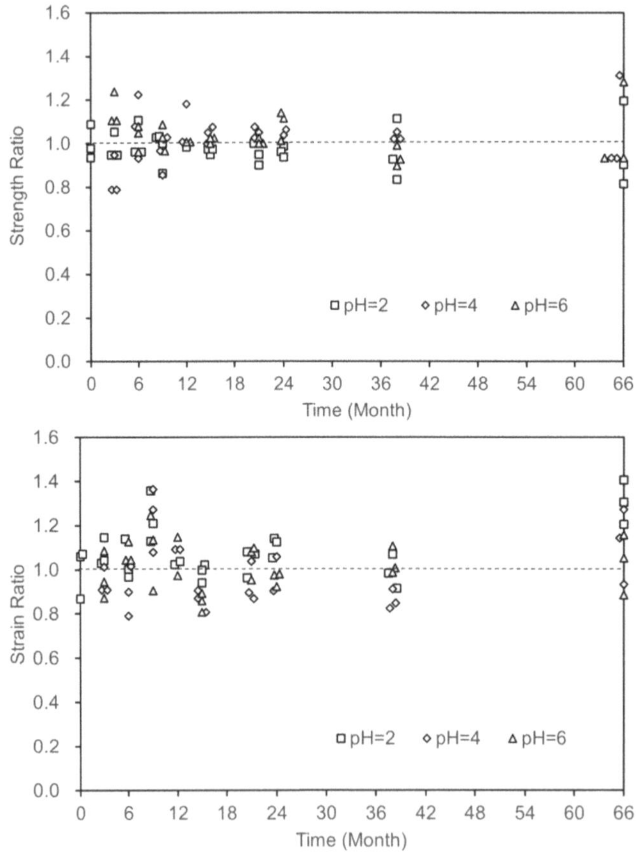

Figure 5.7 Strength (a) and strain (b) ratios for hydraulic asphalt concrete (HAC) specimens in acidic water (pH = 2, 4, and 6) at 20°C (Feng et al., 2020b).

ratio is defined as the tensile strength of the specimen immersed in acidic water divided by the tensile strength of the specimen immersed in neutral water. A ratio less than one indicates that the acidic water has caused some tensile strength reduction. Figure 5.7 also shows the corresponding ratio between tensile strain up to when the specimen failed in tension by splitting.

5.6.2 Conclusions from test program with acidic water

Marshall specimens of HAC were immersed in acidic water for up to 5.5 years. The acidity of the water in the various tests was pH = 2, 4, and 6, and the temperature 20°C. The specimens were subjected to indirect tensile tests (splitting tests) at regular intervals during the testing program.

The test results show no indication of tensile strength reduction with time and with acidity of water, and no reduction in tensile strain up to when the tension crack developed. Based on the test results, one may conclude that any degradation of HAC in acidic water, if it is occurring, is extremely slow. It appears that HAC may safely be used in impervious facings or in cores in tailings dams even for projects requiring long-term storage (containment) of very acidic water. This study included indirect tensile (splitting) tests but no compression or bending tests that should be included to investigate any effects of acidic water on hydro-mechanical properties of HAC in general.

Chapter 6

Core construction and quality control

ICOLD (2018) gives a detailed presentation of hydraulic asphalt concrete (HAC) production and the development of core placement equipment and compaction procedures since the first asphalt core embankment dams (ACEDs) were built. A summary of present practice and quality control procedures and requirements is presented in this chapter.

6.1 Rock aggregate production for core and transition zones

Basically, there are two kinds of aggregates used for making the asphalt concrete core and the transition zones; rock aggregates are obtained by blasting and crushing and by naturally occurring gravel and sand. The shape and gradation of crushed aggregates are strongly affected by the rock types and the type and operation of the crushers. The rock crushing and screening plant must produce aggregates with a gradation that meets the specifications for the core itself and for the granular transition zone on both sides of the core. Normally, the aggregate production for the fine transition zones and the asphalt core is separated as the required gradations are quite different.

The gradation for the fine transition zones (Zones 2) is commonly required to have maximum grain size of about 60 mm and should preferably comply with $d_{50} \geqq 10$ mm and $d_{15} \leqq 10$ mm. The weight of fines (particle size less than 0.075 mm) should not exceed 5% of the total weight of the material in the fine transition zones. The gradation of the aggregates in the asphalt core is usually required to have a maximum grain size of 16–19 mm and comply with the modified Fuller gradation curve as described in Chapter 3. If the gradation of the aggregates for the fine transition zones produced by the crusher has less fine aggregates than required, specially prepared fine aggregates must be added to make the final gradation meet the requirements. Commonly, crushed rock particles with a size between 20 and 40 mm are used for re-crushing and screening for the aggregates used for the asphalt core mix. It is often required to undertake special aggregate production of a particle size less than 2.36 mm as the asphalt core HAC mix requires a high content of fine aggregates.

If natural gravel and sand with the required quality are available, they can be crushed or screened for the aggregates used for the asphalt core. If the natural gravel and sand have the required gradation, they can be screened and used without being crushed. However, natural sand and gravel with sub-rounded particles have lower angle of shearing resistance than crushed, angular aggregates. As mentioned in Section 2.3, for ACEDs higher than about 80 m, some designers prefer to use crushed aggregates, or crushed aggregates should constitute at least 50% of the coarse aggregates for use in the asphalt concrete core.

DOI: 10.1201/9781315177786-6

Natural sand can be combined with crushed coarse and fine aggregates to make up the combined aggregate gradation for the asphalt core. Natural sand with sub-rounded particles makes the mix workable and promotes flexible and ductile stress-strain behaviour. However, in many cases, the clay content is high in a natural sand deposit, and the sand must be washed to get a low clay content, for use in the asphalt concrete mix (see Section 2.3).

6.2 Asphalt concrete production

In some cases, a suitable asphalt concrete production plant may already exist within an acceptable distance from the dam site, and the asphalt concrete may be transported to the site as specified in Section 6.3.

However, if not already available, an asphalt production plant must be built in a location as near as possible to the dam considering the topographic conditions to shorten the asphalt transportation distance and time and not to interfere with other construction activities. The selected location should have good drainage conditions, avoid the threat of flooding, and the asphalt production must meet local requirements for environmental protection.

The time it takes for mixing aggregates and filler is normally 15 seconds and normally 45 seconds for mixing in bitumen. The asphalt concrete mix should be uniform, and all the aggregates should be covered with bitumen. It should be noted that the production capacity of a plant for HAC is typically less than 75% of that for producing asphalt concrete for pavement construction.

ICOLD (2018) gives the following recommendations:

> The asphalt concrete material shall be produced in an asphalt batch mixing plant with capacity to produce a volume at least sufficient for placing up to three layers per day at any level of the dam. The proportions of aggregates, fillers and bitumen shall be added automatically and by weight after hot screening of the various components. The plant shall have an automatic warning or stop by any faulty operation or in-weighing and have an automatic log and print-out possibility for each batch also showing the actual mixing temperature.

Note: The recommended production capacity mentioned above depends on the local situation, the length of the core, and the number of paving machines (pavers) engaged in the same dam construction.

The asphalt plant has separate systems for bitumen, aggregate, and filler, respectively. A typical production flow diagram is shown in Figure 6.1. Bitumen can be either kept in barrels or in bags or canned in a petroleum refinery and transported to the dam site. The way to transport bitumen from the refinery to the dam site depends on the requirements for quality control and economy.

The bitumen at an asphalt plant should be indirectly heated to be dehydrated at a temperature of 110°C–130°C and at a temperature of 155°C–165°C for mixing aggregates. The temperature of the mixed asphalt from the plant is normally between 155°C and 170°C. This is for a mix with bitumen grade B70. Higher grades of bitumen require somewhat lower batching temperatures as discussed in Section 6.2. The bitumen must be weighed separately before mixing with the aggregates, as indicated in Figure 6.1.

Crushed aggregates are transported from the crushing and screening plant to the asphalt plant and should be kept in separate storage bins, one for each different particle size group. The storage bins should have storage volumes to be able to keep the plant continually working for at least

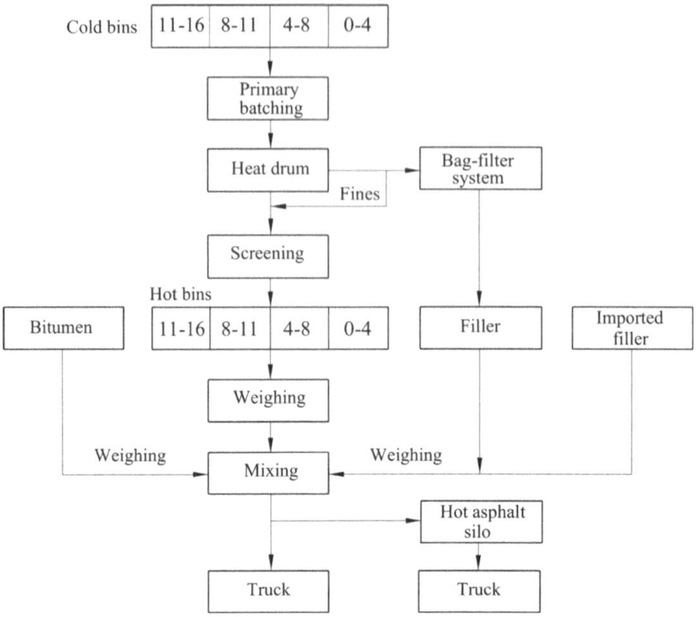

Figure 6.1 Typical asphalt production flow in an asphalt plant.

10 days even if new supplies are not provided during that period. The storage bins at the plant should be roofed against rainy and windy weather. While the plant is working, the aggregates in the storage bins are transferred to the cold bins of the plant. The asphalt plant should have a minimum of 4 (four) cold bins. If natural sand is used as part of the fine aggregates, a fifth cold storage bin is needed. The aggregates in the cold bins are weighed in a proportion according to the asphalt mix design and transferred to the heat drum by a conveyor belt. After having been heated, the aggregate mixture is screened and separately stored in hot bins for different grain particle groups, as shown in Figure 6.1. During the heating, some fine aggregates and most of the filler size material in the mixture in the heat drum are sucked out by the bag-filter system. A part of the fine aggregate and filler fall into inverted cone silos and are transferred to the hot mixture before screening. Most of the filler in the mixture is collected in a silo as retrieved aggregate filler.

Aggregates stored in the different hot bins according to the different particle size groups are separately and cumulatively weighed in the bucket according to the design proportions. The doors of the hot bins are opened one by one, and the aggregates are freely falling into the bucket and being weighed. There are always some aggregates in the "air" after having closed each of the doors. Therefore, the sizes of the doors in the bins, the speeds of the aggregates falling, and the door closing speeds affect the weighing accuracy. The weighed aggregates fall in the mixer from the bucket and are mixed with separately weighed filler and bitumen.

The weighing system must be calibrated regularly to always ensure having the correct weights and proportions. The total filler usually consists of the remaining filler in the fine aggregate (0–4 mm), retrieved aggregate filler during production, and imported filler from an external source. The plant must have a weighing accuracy of $\pm 0.5\%$ for the cumulated aggregates from the hot bins and for the filler that may consist of retrieved aggregate filler and imported filler. The remaining filler content in the fine aggregate hot bin varies and must be checked regularly. The accuracy of the bitumen weight should be within $\pm 0.3\%$.

Figure 6.2 Typical layout of conventional asphalt production plant.

At all times, the temperatures of the aggregates, the bitumen, and the final mix must be automatically controlled. A computer at the plant must show and continually record the proportions for each batch. The computer alerts the operator if the proportion of any material deviates from the specified limits. Scales and screens in the asphalt plant must be checked regularly as errors in these are not detected by the computer. The hot asphalt mix is discharged into a truck or stored in a hot, insulated silo when continuous batching is needed for the plant. The asphalt in the silo should not be stored for more than 8 hours before discharging.

Figure 6.2 shows a typical asphalt production plant arrangement. Traditionally, the bag-filter system and bitumen containers (tanks) have been separated from the hot silos. A large area has usually been occupied by these facilities until recently when a more integrated asphalt plant was developed by Xi'an Lutai Machinery Manufacturing Co., Ltd (Figure 6.3). The integrated plant has the bag-filter system and bitumen tanks in one building. This significantly reduces the area required for the asphalt plant and the required energy supply.

6.3 Transportation of hot asphalt

The hot asphalt is transported from the production plant on to the dam surface by a group of dump trucks to maintain a continuous asphalt core placing operation. In windy, rainy, and cold seasons, the trucks must be waterproofed and insulated to maintain the required temperature of the asphalt concrete. The time required for transporting the asphalt from the production plant to where the asphalt concrete is to be placed in the core should not be more than ½ hour. It is essential that the temperature of the asphalt concrete is always maintained within the specified limits. Furthermore, segregation of the asphalt mix during transportation must be avoided to maintain a uniform mix.

At the dam, an excavator may be used to take asphalt from the truck and transfer the asphalt either directly to the core for hand placement or into a paver for machine placement. A movable

Figure 6.3 New integrated layout of asphalt production plant.

steel framed structure is used to raise the truck to facilitate the unloading of asphalt from the truck into the modified hopper of a reloader. The asphalt in the modified hopper of a reloader is transferred either directly to the core for hand placement or into a paver for machine placement. The hopper of a reloader for transferring asphalt may have to be specially insulated to keep the asphalt hot when unloaded into the hopper.

Due to the limited volumes of the paver silo and the hopper for the reloader, four to five small dump trucks are usually required to transport the hot asphalt mixture from the asphalt plant to the dam site to maintain continuous construction (raising) of the asphalt core. For the Suwalong Dam in Sichuan Province, China, a movable, large asphalt storage silo was built (Figure 6.4). Large dump trucks are used to transport the hot asphalt mixture from the asphalt plant and to unload into the large silo on site.

6.4 Asphalt placement in core

The details of the design of the asphalt core and the foundation plinth are presented in Chapter 8. The design of the concrete plinth at the base of the asphalt core depends on the geological and topographical conditions of the dam site and should be planned such that required hand placement of asphalt concrete is minimized. As the topographical contour of the dam site can be very irregular, some hand placement becomes unavoidable at the bottom of the core and towards the abutments where the placing machine ("paver") cannot reach.

The concrete plinth surface should be given a rough texture for good bonding with the asphalt concrete. Any materials from the injection grouting under the plinth must be removed, and the plinth surface must be cleaned and dried. The plinth surface is normally sprayed with a special coating (primer) in a quantity of about 0.2 kg/m^2. The coating is a mixture of bitumen with some gasoline to facilitate the spraying and the absorption by the concrete surface of the plinth (Figure 6.5).

Figure 6.4 Asphalt concrete is unloaded into the silo from a dump truck.

After the plinth surface has dried and the gasoline in the coating has been fully volatilized, a layer of asphalt mastic with a thickness of about 20 mm is placed to improve (and to make watertight) the connection between the concrete plinth and the asphalt core (Figure 6.6). The asphalt mastic consists of bitumen, filler, and sand and is also called sandy asphalt mastic (SAM). When the concrete plinth is on a steep abutment, bitumen with added styrene-butadiene-styrene (SBS) may be used in the mastic to increase the shearing and tension resistance of the core-abutment interface (Section 3.10). Special investigations of the asphalt-SAM-concrete connection have been undertaken and described by Wang et al. (2018).

Hand placement of the bottom layer of asphalt concrete core takes place inside formwork with transition Zone 2 on the outside. Zone 2 must provide a horizontal and stable base for the core paver as it travels along the centreline of the core (Figure 6.7). After the formwork has been erected, the transition zone material on either sides of the core is placed either before or after placing the asphalt concrete inside the formwork. When the transition zone material is placed before placing asphalt concrete, the core should be covered to avoid any contamination of the previous asphalt layer.

When a horizontal core base of minimum 40–50 m length has been established by hand placement, the asphalt core and transition zone material on either sides of the core are placed by the machine paver (Figure 6.8). However, manual work is required when placing the core towards steep abutments where the paver and rollers for compaction cannot reach. Asphalt concrete against the abutments should be carefully compacted by means of vibratory rammers, tampers,

Figure 6.5 Bitumen coating sprayed on the surface of the concrete plinth after water stops are installed in the plinth.

or plates. Asphalt core placing speed by the paver should be continuously adjusted in accordance with the plant and transportation facilities. The speed is normally between 1 and 3 m/minute. If stops exceed 20 minutes, a transverse construction joint must be made before proceeding with the next layer. The joint is constructed with a slope of 1V:3H.

There are several designs of asphalt core pavers developed over the years by various construction companies such as German Strabag, Norwegian Veidekke, Swiss Walo, and Chinese Xi'an Huize.

The paver is hydraulically driven, and the widths and elevations of the screens for the core and Zone 2 are adjusted according to the design specifications. The elevation of the screen for

Figure 6.6 Sandy asphalt mastic (SAM) manually placed on the surface of the concrete plinth.

Zone 2 is higher than for the core to achieve the same top level of the core and transition zones after roller compaction is completed.

The precise centre of the core is marked for each layer with a line. A video camera mounted in front of the paver and a television monitor inside the cab enable the operator to steer the machine with precision following the core centreline.

ICOLD (2018) recommends that: "To achieve a proper bond between the individual layers, the previous layer shall be reheated with an infrared heater to remove moisture. Additionally, the pre-heating shall be done at least before the first layer of each working day and always when

Figure 6.7 Hand placement of bottom layer of asphalt core inside formwork

climatic conditions or minor and short precipitation periods require this". The main objective of the pre-heating is to ensure that no moisture remains on the previously placed asphalt concrete layer when the new layer is being placed. A secondary objective is to heat the top of the previously placed asphalt concrete layer. The heat contained in the newly placed 0.20–0.30 m thick layer will normally warm up the top surface of the previous layer sufficiently for the layers to melt together. The minimum temperature requirement for the previous layer and the minimum temperature requirement for compaction of the new layer must be ensured. In warm weather, and when the asphalt core is continuously being placed, the surface of the previous layer may not need to be heated by the infrared heater. The new layer increases the surface temperature and makes a satisfactory connection between the previous and new layer after compaction. No tack coat is required between the asphalt layers. Core sampling has proved that the connection at the interface between compacted layers is tight and hardly detectable in specimens drilled through the interface (see Section 6.9.3, Figure 6.17b).

Figure 6.8 Materials in core and Zone 2 simultaneously placed by the paver.

The asphalt concrete and transition zone material (Zone 2) on both sides of the core simultaneously flow out of the paver with the transition zones a few centimetres higher than the asphalt core level, corresponding to the difference in compressibility. The transition zones give the hot asphalt core immediate lateral support. The height difference allowed between the transition zones and the asphalt core depends on the compressibility of the materials in the transition zones and must be tested and verified at the test section at site (see Section 6.6).

In 2015, the Swiss company WALO developed a new type of paver based on the experience from previous projects with various earlier pavers for ACEDs (Figure 6.9). The paver combines all the important parts of the machine in one compact unit. A solid carriage supports a covered and heated asphalt bin, spacious hoppers for transition zone material, an infrared-heating device, and high-duty compaction units. The paver is GPS-controlled in three directions. In combination with the four independent crawlers, the steering mechanism of the paver allows following the central line precisely even for a curved dam axis with small radius of curvature (Figure 6.9). Due to the 3-dimensional GPS-control, the actual paving height can be kept fixed. Several sensors and their recorded data enable consistent and traceable quality control, but continuous visual observations are still important.

The simultaneous placing of asphalt and transition zone material by the paver requires a continuous supply of asphalt concrete and transition zone material to the paver. The transition zone material may in advance be placed in small stockpiles on one side of the core and then, when needed, be transferred into the paver by an excavator (Figure 6.10).

Another method is using a bigger storage bin that is pulled by an excavator, and the transition zone material can be loaded into the bin from trucks (Figure 6.11). This makes the dam surface free of the stockpiles of the transition zone material, but trucks are required to timely transport the material to the dam surface during the paver operation.

Figure 6.9 In 2015, the company WALO developed a new type of paver. (Courtesy of WALO International AG-Department-Hydraulic Asphalt Engineering)

Transport across the asphalt core is normally a necessity for transport of dam construction materials and equipment. Light steel bridges, which can be easily removed, are required, and no vehicle is permitted to cross the core except over such a bridge (Figure 6.12).

6.5 Field compaction procedures

After the material in the core and transition zones (Zones 2) on both sides of the core have been placed manually or by a paver, compaction is achieved by three vibratory rollers. If it is the first layer placed at the start of the day, a 10-minute waiting period before compacting the asphalt layer may be needed to allow the temperature of the surface of the existing layer to be raised by the heat from the newly placed layer. Some pavers are equipped to perform preliminary light compaction by tampers of the asphalt core and transition zones at this stage of the operation. However, the final compaction to obtain the specified densities must be done by rollers as described below.

The weight of the vibratory roller for compacting the asphalt core is normally between 1.5 and 2.5 tonnes with a roller width between 80 and 120 cm. The selection of vibratory rollers for the asphalt core depends on the properties of the asphalt mix and core width. The roller for compacting the core should be somewhat wider than the core itself. However, the roller should not be too wide as then the roller could span over the core between the transition zones without compacting the core.

Figure 6.10 Transition Zone 2 material may be placed in small stockpiles along one side of the core ready for use during core construction.

Figure 6.11 Transition zone material is loaded from trucks into the storage bin and then transferred into the paver by an excavator. (Courtesy of WALO International AG-Department-Hydraulic Asphalt Engineering.)

Figure 6.12 Steel frame bridge spanning over the core, and truck crossing the bridge without loading the core.

For a more detailed discussion of the compaction of the core and transition zones, the presentation by ICOLD (2018) may be consulted. The weight of the vibratory roller used for compacting Zone 2 is normally between 2.0 and 3.0 tonnes with a width of 120 cm. It should be noted that various types of rollers have quite different compaction energy outputs even though their sizes and weights may be the same.

The thickness of the asphalt core layer as placed is commonly about 3 cm thicker than the final layer thickness after compaction. After compaction, the usual layer thickness in previous ACEDs has been between 20 and 25 cm. However, there are also projects where the compacted core layer thickness has been as much as 30 cm with very satisfactory results (see further discussion of rate of core construction in Section 6.8). To facilitate compaction, the gravel material in Zone 2 is usually wetted before being placed. Dry density and porosity are commonly taken as quality control parameters after compaction.

There are two different compaction procedures (A and B) used in practice to compact the asphalt core and the fine transition zone materials. If properly executed, both procedures give satisfactory results. The equipment used is the same in both compaction procedures, but the sequence of compacting the core and the transition zones is different. In compaction procedure A, the asphalt layer is compacted first by the roller using two passes without vibrations and then one or two passes with vibrations to reach a preliminary density (Figure 6.13). This causes the core width to increase a little towards the top of the layer. Then, the transition zone material on

Figure 6.13 In compaction procedure A, the asphalt core is compacted first.

Figure 6.14 In compaction procedure B, the transition zones are compacted first to give the core preliminary early confinement. The photo shows light rollers of small diameter that may be used for the preliminary compaction.

both sides of the core is compacted simultaneously by two rollers in parallel using two passes without vibration and one or two passes with vibration to reach a preliminary density. After the asphalt and transition zones have been given their preliminary densities, more passes with vibration are required to obtain the final porosity of maximum 3% for the asphalt core. Afterwards, two passes without vibration are required to provide smooth surfaces of Zone 2 for the paver to travel on when placing subsequent layers.

In compaction procedure B (Figure 6.14), the material in transition Zone 2 is compacted first. Transition zone material on bothe sides of the core is compacted simultaneously by two rollers in parallel using two passes without vibration and two to four passes with vibration to reach a preliminary density. The edges of the roller drum should be kept 15–20 cm from the sides of the asphalt core to avoid squeezing the core thinner than the specified core width. Then, the asphalt core layer is compacted with the confinement of the transition zones on both sides of the core. Three to five roller passes with vibration over the asphalt core are normally required to satisfy the porosity of maximum 3.0%.

To prevent the asphalt from sticking to the roller during the compaction process, water or a mixture of a small amount of diesel oil and water (ratio 1:3) can be used on the roller surface. However, the amount of diesel must be limited to avoid contaminating the surface of the core.

The asphalt core and Zone 2 are normally placed continuously from one abutment to the other. To get the newly placed asphalt layer compacted before the temperature of the asphalt is decreased below the specified limit, the length from one abutment to the other can be divided into short sections (e.g., 30 m) and compacted by rollers going back-and-forth.

The Zones 2 should be compacted to provide stable and uniform lateral support to the asphalt core and vertical support to the paver during core construction. However, the transition zones should not be compacted too much (over-compacted) because the compression modulus for the transition zones becomes higher than the compression modulus for the core after it has cooled down. A very high compression modulus in Zones 2 leads to arching (transfer of stress) across the core and reduced vertical stresses in the core. This is discussed in Section 8.2.

The elevation of the asphalt core and transition Zone 2 shall not, at any time during construction, be more than two layers above or below the adjacent Zone 3. No more than two layers above is to ensure roller stability against overturning and safety of the roller operators during compaction and to ensure required lateral support for the warm asphalt core. No more than two layers below is to avoid the surface of the previously placed and compacted asphalt layer from being contaminated, especially during rainy seasons.

In case the asphalt mixture cannot be continually placed due to temporary break-down of the construction equipment, or sudden heavy rain, the front end of the new layer should be placed and compacted with a slope of 1V:3H to prepare for the joint to be made when the layer is continued later.

If the construction is conducted in cold weather, and especially if the construction rate is 3 or 4 layers placed and compacted per day (see Sections 6.8 and 6.9), fine longitudinal cracks may appear along the centreline of the core due to a slight convex curvature of the core top surface. These fine cracks are of no real concern as they are longitudinal (not transversal), and experience shows that they completely disappear when the next hot layer is placed and compacted.

The compacted asphalt layer is a little wider towards the top than the average core width due to less lateral confinement for the top of the layer than deeper down. The core width must at all levels be as wide or wider than the specified required core width, and this is controlled at regular intervals by excavating small, narrow longitudinal trenches in Zone 2 on both sides of the core. Figure 6.15 shows a typical cross section after compaction for a case with a core design width of 50 cm.

6.6 Test section built before start of dam core construction

Prior to starting the asphalt concrete core construction work on the dam, a test section must be built of the asphalt core including the fine transition zones (Figure 6.16). The construction of this section tests the field behaviour of the preliminary asphalt concrete design mix. Furthermore, it shall demonstrate that the asphalt plant, the construction equipment, and the personnel

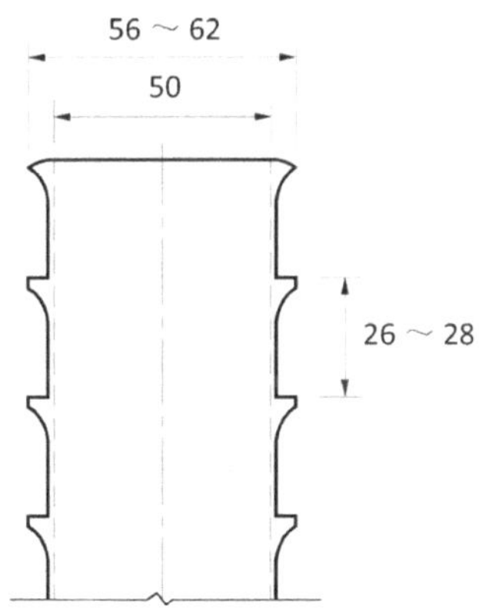

Figure 6.15 Vertical section through the asphalt core with a design width of 50 cm (dimensions given in cm).

(1) Top of the concrete plinth; (2) concrete abutment: (3) copper water stop; (4) formwork; and (5) SAM tank.

Figure 6.16 Full-scale test section. SAM on top of the concrete plinth is being placed and formwork is being installed for preparing the first layer of Zone 2 and asphalt concrete core. The width of the asphalt layer is 70 cm and is flared to 140 cm towards the concrete abutment.

can perform the various dam core construction tasks in a satisfactory manner to fulfil the requirements of the technical specifications.

The program for the test section as recommended by ICOLD (2018) includes:

- Construction of a concrete plinth of 25 m long, 4 m wide, and 0.2 m thick, including installation of water stops.
- Demonstration of the method proposed to clean and to prepare the surface of the concrete plinth.
- Application of mastic over the 25 m plinth length with a width according to design.
- Construction of one layer of asphalt concrete with width according to design and transition Zone 2 hand-placed over the full 25 m.
- Construction of two layers of asphalt concrete core and transition zones, both placed with the core paving machine.

If the design and construction schedule call for three or four asphalt layers to be placed per day (see Section 6.8), it must be demonstrated that the quality of the compacted HAC and the interfaces between the layers are satisfactory. For a compacted asphalt layer thickness larger than 25 cm, it must be demonstrated by construction of the test section that the quality of the HAC is satisfactory over the full height.

The construction of the test section must demonstrate that the plant and equipment to be used for the dam core construction have the following capabilities:

- The asphalt plant has scales with the required weighing accuracy, and the temperatures of aggregates, bitumen, and asphalt mix are controlled to be within the allowable ranges.
- The plant mixing procedures can consistently produce a uniform asphalt mix with an aggregate gradation curve, filler content, and bitumen content within the specifications.
- The selected vibratory rollers, number of passes with and without vibration, and the compaction procedures can compact the asphalt and transition zone layers with the specified thicknesses to meet required densities (porosities). The minimum width of the core and half of the width of the core to the centreline meet the requirements.
- The treatment and temperature control of the surface of the previous layer produce satisfactory bonding between existing and new layer after compaction.
- The selected sandy asphalt mastic (SAM) provides watertight bonding between the asphalt core and the concrete plinth.

In addition, the contractor may be required to demonstrate the following capabilities:

- The asphalt mix may be adjusted and improved due to a change in the aggregate gradation curve by mixing at the asphalt plant and compacting by a roller at the site.
- The placing of SAM on an inclined plinth and placing and compacting the asphalt layer flaring towards the inclined plinth.
- Drilling out (extracting) quality control samples and subsequently filling and compacting in the open drill holes.
- Placing and compacting transition Zone 2 to coordinate with the placing and compaction of the asphalt core.

After the asphalt core in the full-scale test section has cooled down, at least three undisturbed core samples with 100 mm in diameter shall be drilled out (extracted) at each of the following locations:

- The hand-placed layer and the interface between the hand-placed layer and the concrete plinth.
- The interface between paver-placed and hand-placed layer.
- The interface between two paver-placed layers.

The core specimens shall be drilled to a depth of at least two layers (e.g., 0.45 m if compacted layer thickness is 20 cm), and each core shall be logged, inspected, and tested for air void content, aggregate gradation, bitumen content, and filler content. If required, some core specimens are drilled out and sent to an experienced external laboratory for conducting triaxial compression control tests. Bending and tension tests may in some cases also be required to further investigate the hydro-mechanical properties and the cracking resistance of the compacted asphalt concrete (see Section 3.8).

Presently, available geophysical methods are not sufficiently accurate to obtain results for air void content of the compacted asphalt layers without calibration with drilled-out samples. Establish the relationship of the air void content measured on drilled specimens and geophysical readings on the same core drilling points by a nuclear or non-nuclear instrument. That relationship (calibration) must be established before the geophysical measurements can be relied upon to determine asphalt porosity. The physical sampling by drilling 100-mm diameter core specimens is required to establish the calibration of geophysical measurements of the in-situ air void content to be used in the construction quality control.

A comprehensive report on the test section, including all test results and recommendations, shall be submitted for approval before any work commences on the dam core. If the results from the test section are unsatisfactory, a new test section must be constructed.

6.7 Placing and compacting HAC in cold weather

When constructing asphalt core in cold weather, a full-scale test must be done to validate the placing and compaction procedure and the quality of the compacted HAC. The quality requirements must be the same no matter what weather conditions are encountered. Cold weather conditions affect the whole system of constructing asphalt core such as asphalt mixing, transporting, placing, compacting, and protecting the compacted layer.

To facilitate the placing and compacting in cold weather, a bitumen content of about 0.2% points may be added to the design asphalt mix, and the temperature of the mix to be placed may be increased by 5°C–10°C. The lower part of the batching plant may have to be enclosed by canvas to prevent heat from escaping when the hot asphalt mixture is (transferred) into the truck. The bottom of the truck may have to be insulated by a thermal insulation plate, and the truck be covered with canvas. The last asphalt layer placed and compacted for the day must be covered by canvas or a quilt with cotton wadding after compaction is completed.

A full-scale test must be conducted during the anticipated coldest weather conditions at site. The time for performing the test layer is often selected during the night as the temperature at that time is the lowest during the 24-hour period. Air temperature and temperatures of the previous layer and in the newly placed layer must be measured continuously during the full-scale test.

After the test layer is cooled down to ambient temperature, asphalt specimens must be drilled out from the test layer. The quality of asphalt concrete at the interface between the two layers and of the test layer placed at the cold temperature must be as good as that constructed at normal temperatures.

Such full-scale tests have been conducted in low temperatures down to as low as –15°C for several ACEDs, e.g., Yele Dam, Wuyi Dam, Ertanggou Dam, and Momoke Dam, all in China. The asphalt cores in these and other dams that were constructed in cold temperatures have been reviewed and are reported to perform well. Therefore, the procedures followed for these dams are recommended as procedures to be followed when placing asphalt concrete in cores in cold weather.

When an ACED is to be built in a narrow valley, strong channel crosswind may further reduce the temperature of the hot asphalt mixture during construction at the site. The wind speed at the core location can be significantly reduced by raising the upstream and downstream shoulders higher than the core to mitigate the situation (Guo et al., 2020).

In summary: Placing and compacting HAC in cold temperatures lower than –10°C is not recommended. As explained above, that requires significant extra efforts to ensure the quality of the asphalt concrete, to keep the asphalt plant and construction equipment in operation, and the working conditions for the field crew get to be very demanding.

6.8 Asphalt core construction rate

The construction rate for the asphalt core is generally limited by the rate of construction of the outer transition zones and shoulders. Normally, significant parts of the upstream and downstream shoulders are built during dam foundation excavation and grouting. Therefore, the construction rate of the asphalt core can be high in the beginning as the remaining volumes to be placed of the upstream and downstream shoulders are small. When the dam rises, the construction rate of the asphalt core slows down as the rate of height increase for the shoulders is slow. In the top part, the construction rate of the asphalt core increases again as the required volumes of the material in the shoulders decrease.

Traditionally, a construction rate of 2 (two) asphalt layers placed and compacted per day (i.e., a core raise of 0.4–0.5 m/day) was sufficient to meet the normal rate of dam embankment construction. However, after larger and better construction equipment for excavation, transportation, and compaction of rockfill and alluvium has become available, a higher rate of asphalt core construction has been made possible.

Tschernutter (1997) reported that the maximum construction rate of the asphalt core for the Feistritzbach asphalt core rockfill dam in Austria was 3 layers of 20 cm per day (i.e., 60 cm/day). Saxegaard (2002) built a 15-m-long test section and demonstrated that the construction rate of asphalt core could be increased significantly without sacrificing quality. One field test was to place four layers with a compacted thickness of 20 cm/day, and another test was to place three layers with a compacted layer thickness of 30 cm that gave a rate of 90 cm/day. Test results of drilled core specimens showed air void contents meeting the specified requirement of maximum 3% for the two ways of accelerated asphalt core construction, and the interfaces between layers were hardly detectable.

Dams are often required to be raised quickly up to a height sufficient to prevent overtopping by floods during the construction period. Wang et al. (2020a) presented a case study for an ACED in China that demonstrates that the rate of construction can be increased, still providing satisfactory quality of the compacted asphalt concrete. The 63.6-m high asphalt core for the 71-m high Zhaobishan ACED was completed in 159 days with a compacted layer thickness of 23 cm. Among the 159 days of construction, 35 days used the construction rate of 3 layers/day (69 cm/day) and 11 days used the rate of 4 layers/day (92 cm/day). Control tests showed that the requirements of maximum 3% air porosity and the specified core width were also satisfied for the third and fourth layers.

With a modern asphalt plant and asphalt core paver, a construction rate of 4 layers/day is becoming an accepted practice for large ACEDs in China. The asphalt cores of the Shimen and Kayingbulake ACEDs in the Xinjiang Uygur Autonomous Region were constructed at a peak construction rate of 18 m/month. The fact that one can significantly reduce the probability of overtopping during construction by increasing the construction rate is a very important consideration in many situations and affects the safety, economy, and progress of the project.

For the 75-m high Momoke ACED in the Xinjiang Uygur Autonomous Region with a crest length of 370 m, the tight construction schedule became even tighter due to the impact of the COVID pandemic. The asphalt core was constructed at a peak construction rate of 20 m/month (3 layers/day) to reach the flood control elevation in time in 2023.

6.9 Quality assurance and field control specifications

6.9.1 *Validation of asphalt concrete mix*

Aggregates, filler, and bitumen to be used must meet the requirements stated in Chapters 2–4. The asphalt mix design must be carried out and tested in an experienced laboratory. The laboratory must submit a report on the material properties and the hydro-mechanical behaviour of the recommended asphalt mix design.

The dam site conditions and geologic materials available vary from project to project. It is therefore difficult to give "typical" technical specifications for an asphalt mix design. However, general technical specifications can be given for reference and are shown in Table 6.1.

The asphalt mix design, based on laboratory testing, must be validated by investigations at the full-scale test section (see Section 6.6) and continuously during the construction period. Some modifications and adjustments of the preliminary mix are frequently required to provide the

Table 6.1 General technical specifications for a hydraulic asphalt concrete (HAC) mix for core construction

Content	*Item*	*Unit*	*Requirement*	*Note*
Bitumen			Refer to Table 2.1	Bitumen specifications are different from country to country and should be specified by the designer
Coarse aggregate	Apparent density Bitumen-aggregate adhesion Flakiness Index (FI) L. A. abrasion test Durability Water absorption Clay content	g/cm³ Grade % % % %	≧2.6 ≧4 ≦35 ≦45 ≦12 ≦2.0 ≦0.5	Alkaline material is preferred. Water boiling method or rolling bottle method (European Standard) to test bitumen – aggregate adhesion
Fine aggregate	Apparent density Durability Water absorption Clay content	g/cm³ % % %	≧2.55 ≦15 ≦2.0 ≦2.0	For natural sand
Filler	Apparent density Hydrophilic coefficient Water content Passing 0.075 mm size sieve	g/cm³ -- % %	≧2.5 ≦1.0 ≦0.5 ≧85	Alkaline material for filler is required
Impermeability	Air porosity		≦2.0% ≦3.0%	For Standard Marshall Specimen For drilled core sample
Behaviour	Tensile strain Bending strain Triaxial test (shear strain)	% %	≧1.0 to fail ≧2.0 to fail Ductile	Triaxial stress-strain test curves should show no significant strain-weakening for shear strains beyond peak shear stress. Technical specifications may give specific criteria for the project.

asphalt mix finally recommended for construction. Early tests are usually carried out on the asphalt mix using aggregates crushed in the laboratory and using a laboratory compaction method. At a later stage during the project, tests should be carried out on laboratory-prepared samples using aggregates prepared by the field crusher and on drilled asphalt specimens extracted from the test section. After core construction has started, test specimens should be drilled out of the dam core, as such specimens give the best documentation of the constituents of the asphalt mix and the most realistic stress-strain parameters for stress-deformation analyses and sometimes permeability coefficients for seepage analyses.

6.9.2 Quality control of HAC mix

The aggregate gradations in the various hot bins (see Figure 6.1) should be checked frequently to ensure by careful weighing that the asphalt mix satisfies the specified aggregate proportions within acceptable limits.

The examination of the asphalt mix by either the centrifuge extraction method or the combustion furnace method must be done at least once per working day. The extraction results should be compared with the target asphalt mix, and the deviations from the aggregate curve, filler content, and bitumen content should meet the requirements shown by the specifications in Table 6.2. The sampling must take place at roughly regular intervals during the production and at different locations on the core. The samples should be mixed for the mix deviation analyses (Table 6.2 and Section 6.9.5) and for preparing the standard Marshall specimens to check the density (air void content).

All the time during core construction, there should be at least one air void content test (air porosity) and computation per working day using Marshall specimens prepared of asphalt mix taken directly from the asphalt production plant.

6.9.3 Quality control by drilling and extracting samples from the core

Core drill control samples (diameter of usually 100 mm) may be taken when the asphalt temperature has decreased and allows sampling of undisturbed samples. The most critical part of the sampling process is shearing off the drill sample by rotation at the bottom end. A special technique is used to minimize specimen disturbance: After having drilled down to reach the specified depth, the specimen can be pulled out by installing a thin steel wire loop around the bottom end of the specimen.

To ensure that the quality of the constructed asphalt core meets the design requirements, drilled core specimens should be taken as soon as the first 2 (two) asphalt layers have been placed by hand and compacted to determine the air void content, aggregate gradation, filler and bitumen contents, and the interface conditions between compacted layers and between the concrete plinth and the bottom layer. After having ascertained satisfactory control results, core

Table 6.2 Allowable asphalt mix deviations

Material	*Coarse aggregate (>2.36 mm)*	*Fine aggregate (2.36–0.075 mm)*	*Filler (<0.075 mm)*	*Bitumen*
Deviation (%)	±5.0	±4.0	±2.0	±0.3

construction may proceed using the core machine paver. Figure 6.17 shows examples of drilled asphalt core specimens with a diameter of 100 mm and a cut height of 60 mm. The figure also shows a longitudinal section through the axis of an about 420-mm long core where no interfaces between the layers can be observed. The asphalt content (B70) for the HAC mix in this case was 6.4% and the air porosity less than 2.0%.

(a)

(b)

(a) Cross sections of drilled core specimens. [Some local air voids (black spots) may be detected]. Computed air void content is < 2%. The air void content (air porosity) is usually required to be not more than 3% for asphalt concrete compacted by field rollers in a dam core. (b) Longitudinal sections of drilled core specimens

Figure 6.17 Asphalt concrete specimens about 420 mm long with a diameter of 100 mm cut into (a) specimens of 60 mm height to test the air void content of each, and (b) longitudinal section of drilled specimen to note if any interfaces may be observed between compacted layers. The specimens were drilled out from the asphalt core in the Jiangjiakou ACED, China.

During subsequent core construction, after the first two bottom layers have been controlled, and at least three core samples should be drilled out (extracted) after the first 2–3 m increase in core height (or per working week of construction) to check the key parameters, i.e., air void content, aggregate gradation, filler content, and bitumen content. Each core must be drilled through (at least) two layers such that the interface between the layers can be evaluated by visual observation.

Tests on specimens (diameter of 100 mm) drilled out of the compacted dam core are required to ensure asphalt core quality with an air void content of maximum 3.0%. (Air void contents measured by nuclear or non-nuclear geophysical tests are only considered supplementary and are not sufficient by themselves without calibration with drill core specimens). The drilling of control specimens during construction cannot be achieved until the asphalt has cooled down sufficiently. The cooling may take 3–5 days depending on the weather conditions and whether extra cooling measures are implemented. The control sampling may slow down the rate of core construction and must be planned and scheduled carefully to minimize delays.

As construction progresses and if the results all along have satisfied the specified requirements every 2–3 m increase in core height, control core specimens are taken and tested less frequently, for instance, each time there has been a 6–8 m increase in height. The field control team for the project gains experience with time, gets familiar with the mix and its field behaviour, and can then better judge the quality of the mix and the compacted core without frequent control sampling. Throughout the process of placement and compaction, there must be continuous temperature control to ensure that the temperature is within the specified upper and lower limits. There are special requirements when core construction takes place in cold weather as described in Section 6.7.

The specified frequency of core sampling depends somewhat on the expertise of the contractor building the core and the properties and characteristics of the asphalt concrete mix. The sampling and control frequency must be defined in the technical specifications for the project and may be somewhat different for different projects.

The boreholes left by the drilled core samples must be heated and then filled with hot asphalt in layers approximately 50 mm thick and compacted layer by layer by a hammer like the Marshall compaction hammer (tamper) used in the laboratory investigations or by a vibratory hammer.

6.9.4 Geophysical control measurements of asphalt density

There must be observations of the asphalt by an experienced team during the whole process of mixing, transportation, placement, and compaction. Removal of unacceptable asphalt placed in the core is difficult, time consuming, and costly, especially if the asphalt concrete has hardened before the removal starts. The number one rule to follow is therefore to ensure the quality of the asphalt delivered on the dam for placement and compaction in the core.

After compaction, each layer should be checked by means of a geophysical, non-destructive test method (by a nuclear or a non-nuclear instrument) at different points along the core to calculate the air void content by measuring the in-situ density. The instrument (e.g., Troxler 2710B) is using a spiral magnetic field to measure the permittivity of asphalt concrete. The density and permittivity of asphalt concrete are proportional. After the calibration has been made between the measurements given by the geophysical instrument and known densities (air void contents) of asphalt specimens, the instrument may be used in the field to inspect the density of the compacted asphalt layer in the core. However, until the geophysical measurements have been further developed by state-of-the-art improvements to give more accurate measurements, they must be supplemented and calibrated against physical measurements on drilled and extracted field core specimens.

The geophysical instrument (e.g., Troxler 2710B) can be used to a depth of 5–10 cm from the asphalt surface with an accuracy of 0.003 g/cm³. To the best of the authors' knowledge, there is no available instrument to measure the density (air porosity) in the asphalt core beyond that depth.

6.9.5 Measurements of density deviations – effect of volume of control sample

If the calculated air void content, based on the samples drilled out of the dam core, is found to be higher than 3.0%, the question arises whether the asphalt concrete must be removed, or corrective mitigating measures must be applied. The measurements and calculations should be carefully examined and checked because core material removal may require significant efforts and time if the asphalt layer has hardened.

The densities of coarse aggregates, fine aggregates, filler, and bitumen are required as inputs in the porosity calculations. The air porosity is calculated by the following weight measurements and Eqs. (6.1) and (6.2):

$$\alpha = \left(1 - \frac{\rho}{\rho_{\max}}\right) \times 100\% \tag{6.1}$$

$$\rho_{\max} = \frac{m}{\dfrac{m_c}{\rho_c} + \dfrac{m_s}{\rho_s} + \dfrac{m_f}{\rho_f} + \dfrac{m_b}{\rho_b}} \tag{6.2}$$

where,

α = air porosity of asphalt concrete, %;
ρ = density of asphalt concrete, g/cm³;
$\rho_{\max}$ = maximum density of asphalt concrete, g/cm³;
$m = m_c + m_s + m_f + m_b$, weight of asphalt concrete, g;
m_c, m_s, m_f, m_b = weights of coarse aggregate, fine aggregates, filler, and bitumen, respectively, g;
ρ_c, ρ_s, ρ_f, ρ_b = densities of coarse aggregate, fine aggregates, filler, and bitumen, respectively, g/cm³.

There are basically two methods to determine air porosity of compacted asphalt samples. One method is simple but less accurate with some deviations while the other method is accurate but more time-consuming. The simple method is to calculate the maximum density of the asphalt sample for a given asphalt mix recipe (Eq. 6.2). The more time-consuming method is to directly measure the maximum density of the asphalt sample using a solvent to dissolve the bitumen in the asphalt sample which creates harmful fumes.

For the simple method, the variations in the calculated air porosity depend on the volume sizes of the asphalt samples used for the measurements. As an example, the relationship between variations in calculated air porosity and sample volume size was investigated for a specific case. The densities of the coarse aggregates (2.36–19.0 mm), fine aggregates (0.075–2.36 mm), filler (<0.075 mm), and bitumen were measured for a typical asphalt concrete mix. The maximum density of the asphalt concrete mix was calculated using Eq. (6.2). The weights of coarse aggregate, fine aggregate, filler, and bitumen were weighed with a precision of 0.1 g according to the weight ratio of the asphalt mixture recipe for three asphalt concrete blocks each with a volume of about 3,500 cm^3. The asphalt concrete blocks were prepared using the SFC method (see Section 4.5). The densities of the asphalt blocks were determined by measuring the weight and volume of the blocks. The blocks were cut into half, to 1/2 of original volume of the blocks, and the densities and volumes of the 1/2 block samples were measured. The same was done for the 1/4, 1/8, and 1/16 block samples. Based on the measured densities for all the samples and using the maximum density for the asphalt mix, the relationship between the calculated air porosity and volume of asphalt samples was established. The test results show that the deviations of density become larger when the volume of the samples is less than 1,000 cm^3. The smaller the volume of the samples, the larger the deviation. The largest deviations in density and in turn in the calculated porosity were caused by the large deviations in the asphalt recipes.

Statistically, for the specimens with a diameter of 100 mm, the deviation of the calculated air porosity is ±0.50% for the size of the standard Marshall specimen (63.5 mm in height), ±0.20% for the size of the compression specimen (100 mm in height), and ±0.15% for the size of the triaxial specimen (200 mm in height). Such variations for the calculated porosity should be considered in quality control evaluations during construction and discussions whether the measured air porosity is acceptable.

6.9.6 Controlling width of asphalt core during construction

The width of the core after compaction and cooling must be controlled periodically after careful local removal of the gravel in narrow trenches in the fine transition zones (Zone 2) on either sides of the core. The core width shall not at any location be less than the specified theoretical width, and not less than half of the specified width should be on both sides of the core centre line. Accordingly, a certain amount of overconsumption of asphalt concrete must be anticipated (Figure 6.15).

6.9.7 Insulating core surface during construction shut-down periods

If construction of the asphalt core has to be stopped for some time due to cold winter weather, the core top surface should be covered by insulating materials, e.g., canvas plus a 1–3 m thick layer of transition Zone 2 material. The thickness of cover depends on local freezing depth. When construction of the asphalt core is re-started after the cold period, the temperature in the top of the core is near the environmental temperature. The insulating materials should be removed, and hot aggregates are placed on the canvas to heat the top of the core and increase the temperature in the asphalt gradually and evenly (Figure 6.18). This is a practical method that avoids overheating of the top of the core and thereby "over-ageing" of the asphalt concrete.

6.9.8 Summary of core quality control requirements during construction

- After compaction, the air void content of each new asphalt concrete layer should be checked by means of a geophysical test method (using a nuclear or a non-nuclear instrument) at different points along the core with approximately constant intervals.
- Throughout the core construction period, there should be at least one air void content computation per working day on compacted standard Marshall specimens prepared of asphalt mix

Figure 6.18 After the cold winter intermission, the top of the core may be heated by placing hot aggregates on a layer of canvas to increase the temperature gradually and evenly.

taken directly from the asphalt production plant. Allowable mix deviations (given in % of mineral weight) are presented in Table 6.2.

- Three (3) control specimens drilled out of the core should be taken as soon as the first 2 (two) asphalt layers are placed and compacted in the dam core to check the key parameters: air void content, aggregate gradation, filler and bitumen contents, and the interface conditions between compacted layers and between the concrete plinth and the bottom layer of the core.
- During subsequent core construction, at least 3 (three) specimens should be drilled out after each 2–3 m increase in core height (or per working week of construction) to check the air void content, aggregate gradation, and filler and bitumen contents. Each specimen must be drilled through at least two compacted layers such that the interface between layers can be evaluated by visual inspection. The drilling of specimens to control in-situ air void content cannot be achieved until the asphalt has cooled down sufficiently. The cooling may take 3–5 days depending on weather conditions and whether extra cooling measures are implemented. The drilling and sampling of the specimens must not cause specimen disturbance that may affect the density control measurements or the stress-strain curve of the specimen in a triaxial test. The use of a wire slipped under the specimen bottom may be used to assist with the undisturbed sampling.
- As construction of the core progresses and if the results of the previous controls at 2–3 m height intervals have satisfied the requirements, control samples may be taken less frequently, for instance each 6–8 m increase in core height. Throughout the construction process, there must be continuous asphalt concrete mix and temperature controls to ensure that the mix and temperature are within the specified upper and lower limits during the placing and compaction of new layers.
- The open boreholes left by the drilled core samples must be heated and then filled with hot asphalt concrete in layers of approximately 50 mm thickness and compacted layer by layer by a tamper like the Marshall hammer used in the laboratory investigations.

6.10 Potential future improvements in dam core construction

In present practice, bitumen and aggregates must be heated to high temperatures to be able to mix and compact asphalt concrete to reach the required density (i.e., air void content maximum 3%). The bitumen is unavoidably "aged" (i.e., hardened with reduction in penetration value) by the heating and the duration that the asphalt stays at high temperatures. Innovative construction techniques have been studied to look for ways to make asphalt concrete with the desired hydro-mechanical properties without heating to high temperatures and to facilitate the construction.

Shoichi et al. (2012) reported that foamed asphalt was used for constructing the 15-cm thick base layer of the upstream asphalt facing (area of 178,000 m^2) for the upper reservoir at the Kyogoku Project. The void content was required to be less than 20% for the foamed asphalt base layer. However, foamed asphalt has not yet been attempted for asphalt core construction.

Another potential construction method is to use emulsified bitumen mixing aggregates at normal temperature. The principle is to blend water at a temperature of about 80°C with bitumen at a temperature of about 150°C to form emulsified bitumen with the aid of emulsifier and stabilizer. Emulsified bitumen has very low viscosity because it contains about 50% of water and can easily be mixed with aggregates at normal temperature. Emulsified asphalt has been used for repairing asphalt pavements as the de-emulsification can be conducted by using microwave irradiation technology to form a stiff asphalt pavement during compaction. Emulsified asphalt can be easily dehydrated after de-emulsification as asphalt pavement has a porosity of 3%–10%. However, emulsified asphalt is difficult to use as dense impervious asphalt in a dam core as it is difficult to squeeze water out of the emulsified asphalt during compaction to reach the requirement of air porosity not more than 3% for the asphalt concrete barrier in embankment dams. The key challenge is to develop technology to reduce the viscosity of bitumen to make asphalt that can be easily mixed and compacted and recover the original high viscosity of bitumen after the dense asphalt has been put in place. Additives have been introduced to blend with the bitumen, and comprehensive investigations have been carried out. The research results are encouraging but the developed technology has not yet been put into engineering practice.

Chapter 7

Stress-strain-time-temperature behaviour of HAC

7.1 Loading and deformations of the dam core

Core deformations accumulate with time due to viscous flow of the asphalt concrete, increasing gravity loads during construction, compaction of Zone 2, impounding of the reservoir, seasonal lowering and raising of the reservoir water level, creep in the dam shoulders and transition zones, and earthquake loading for dams in seismic regions. The deformations that occur while the core is still relatively hot may be accommodated by the core without cracking. After the compaction of a new layer, the temperature in the asphalt core cools down and is governed by the temperature in the reservoir water and by the surrounding air temperature as the asphalt concrete gradually acquires its long-term hydro-mechanical properties.

If the embankment is to be founded on a compressible foundation or a foundation with greatly varying compressibility or depth to bedrock, differential distortions due to unequal settlements under the embankment are likely to occur both across and along the valley that may significantly affect the settlements of the plinth under the core and shear distortions in the asphalt core. To avoid these settlements, the plinth may be placed in a deep trench down to firm ground or bedrock, or the plinth is designed to sit on top of a cut-off wall (see Section 8.5).

Asphalt concrete is a time- and temperature-dependent viscous, elastic-plastic material. When subjected to additional loads or deformations, the behaviour of the core strongly depends on the applied strain rate (loading rate) and the temperature conditions. The dam core may typically be raised on top of the plinth at a rate of 2–4 compacted layers added per day (i.e., about 0.5–1.0 m core raise per day) as described in Section 6.8. It typically takes three or more construction seasons to build a high embankment dam, and the rate of first impounding depends on the hydrological conditions and on dam safety considerations. The reservoir impounding and the embankment dam construction may occur simultaneously. This allows earlier use of the reservoir, for instance, for electric power production. It is also beneficial for the stress-deformation performance of the core that is allowed to gradually adjust to the slowly increasing water loads during core construction and to undergo deformations including creep without crack formation. However, during construction and operation, the dam may also be subjected to dynamic earthquake loading with high cyclic loading rates.

The asphalt core is positioned in the interior of the dam, protected by embankment shoulders both upstream and downstream, and during operation, it is in a nearly constant temperature environment. The environmental conditions are favourable with virtually no weathering effects of the asphalt concrete in the core. The temperature of the hot asphalt concrete mix is between 140°C and 160°C during placing and compaction. If the reservoir does not start impounding until after end ofconstruction (EoC), field measurements show that the temperature in the asphalt core may take about 6 months to decrease to an almost constant temperature. Field observations

DOI: 10.1201/9781315177786-7

for the Quxue ACED (see Section 10.7) indicate that the temperature in the core dropped from around 150°C to around 40°C during the first 1 month after compaction of the last layer and to around 20°C after half a year (Feng et al., 2020a). The reservoir was impounded after EoC in the case of the Quxue Dam. When a dam is built while the reservoir is simultaneously being impounded, the asphalt core cools down more quickly as was the case at the Storglomvatn and Storvatn ACEDs in Norway and at Yele ACED in China.

One of the main purposes of performing early triaxial tests on trial mixes (suitability testing) of the asphalt concrete is to study the ductility of test specimens after the triaxial specimens have been loaded to maximum (peak) deviator stress. The degree of strain-weakening (strain-softening) for shear strains beyond peak should be investigated. If the specimen shows a high degree of shear dilation and strain-weakening, the mix must be modified. Any one, or a combination of the following methods, may be applied to promote improved ductility: increase the bitumen and filler contents, use a grade of bitumen with a higher penetration value, reduce the flakiness index (FI) of the coarse aggregates in the mix, add more natural sand with sub-rounded particles rather than crushed sand with angular particles, or using additives and reinforcing fibres. When deciding which of the mixes to be used in the dam project, the influence of the method of laboratory compaction used on the recorded stress-strain behaviour in triaxial tests must be considered and accounted for (see Chapter 4).

General constitutive material formulations that account for strain rate and temperature effects have been developed for viscoelastic materials. However, these general formulations and complex constitutive relationships are not convenient for use in parameter studies during the suitability testing and preliminary design of an ACED. When evaluating core deformations and safety margins against detrimental cracking, a simplified approach may be applied.

The stress-strain and strength parameters relevant for various loading rates and temperature conditions are discussed in the following subsections. The parameters are determined based on the testing of asphalt concrete specimens that have undergone a period of creep and have reached the "creep-stable state" as defined in Section 7.4.

7.2 Effects of strain rate and temperature on asphalt concrete behaviour

In Section 3.7, the results of triaxial tests performed at a specified strain rate and temperature are presented. The material parameters derived from the triaxial test results (e.g., compression modulus, shear and bulk modulus, Poisson's ratio, peak shear strength, shear dilation, and behaviour during plastic yielding) are only valid for the given strain rate and for the given test temperature. However, as will be shown, by using a limited number of tests and test results, one can predict the asphalt concrete stress-strain-strength behaviour also at other strain rates and temperatures by employing the time-temperature superposition (TTS) principle and shift functions as explained in Section 7.3.1.

Feng et al. (2020c) investigated the stress-strain-strength behaviour of HAC by conducting triaxial tests at different temperatures in the range of 5°C–60°C and at different axial strain rates in the range of 0.1%–0.00025%/minute. All the asphalt specimens tested were drilled out of the asphalt core of the Laojiaoxi ACED during construction. The bitumen content (B70) was 6.5%. The series of triaxial tests presented here was run at an imposed (strain-controlled) axial strain rate of 0.1%/minute and at various temperatures applying a constant confining stress of 0.8 MPa. Figure 7.1 shows typical curves of deviator stress ($\sigma_1 - \sigma_3$) vs. axial strain for the asphalt specimens at various temperatures. Figure 7.2 shows typical curves of deviator stress vs. axial strain for asphalt specimens all at 20°C and various axial strain rates.

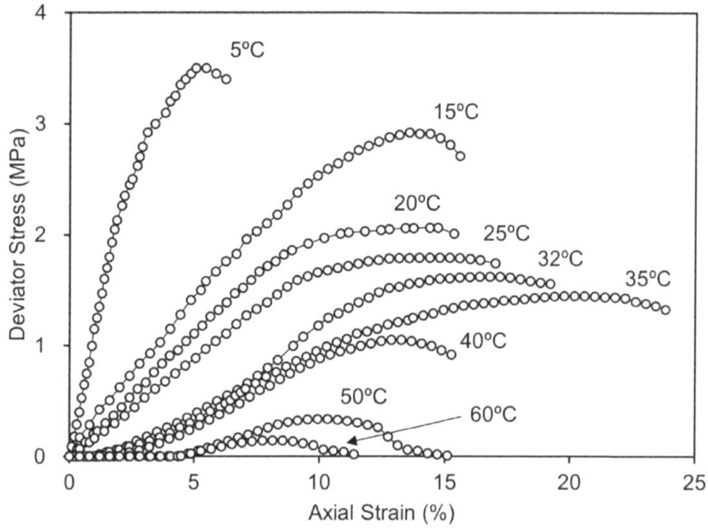

Figure 7.1 Deviator stress vs. axial strain for asphalt specimens at an axial strain rate of 0.1%/minute at various temperatures. Applied constant confining stress 0.8 MPa (Feng et al., 2020c).

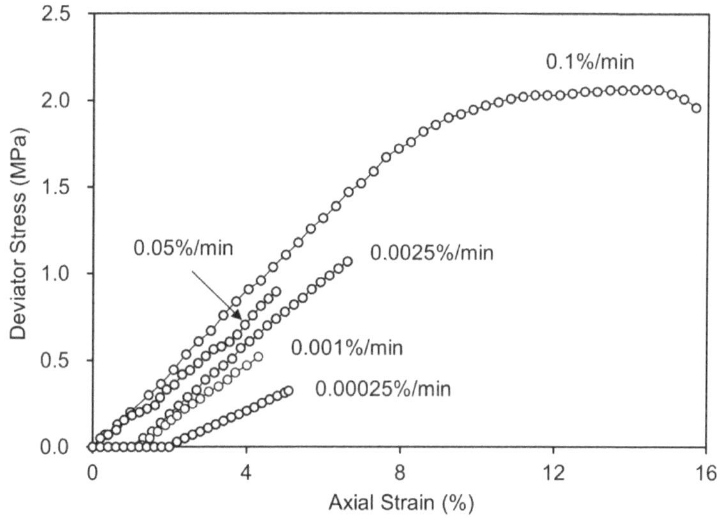

Figure 7.2 Deviator stress vs. axial strain for asphalt specimens at 20°C and various axial strain rates. Applied constant confining stress 0.8 MPa. Tests with the slowest strain rates were terminated when the axial strain reached about 5% (Feng et al., 2020c).

The results from the strain-controlled tests in Figure 7.1 show that when the temperature was 60°C or higher, the measured deviator stress (shear resistance) was virtually zero for axial strains less than about 5%. The test results in Figure 7.2 show that when the axial strain rate was equal to or smaller than 0.0025%/minute, virtually no deviator stress was built up in the initial

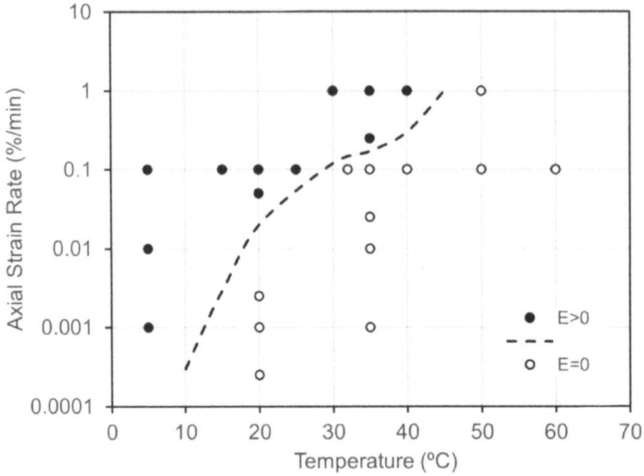

Figure 7.3 Young's modulus *E* (secant up to 1% axial strain) in a graph of axial strain rate vs. temperature for triaxial tests performed by Feng et al. (2020c).

axial strain range. Beyond the initial range, the deviator stress increased linearly with magnitude of axial strain.

The secant modulus up to 1% axial strain was calculated from the test results, and Figure 7.3 shows the results in a graph of axial strain rate vs. temperature.

At any given temperature, asphalt concrete deforms gradually if it is loaded slowly. If loaded at a high strain rate, it is stiff and may even be subject to brittle failure for very high loading rates. At any given strain rate, there is a temperature above which asphalt concrete relaxes quickly enough such that virtually no deviator stress (shear resistance) is mobilized, and plastic flow occurs. Figure 7.3 shows an approximate "boundary" (dashed line) named by Kim (2009) as the stress-free temperature and strain rate boundary below which $E \approx 0$ as virtually no deviator stress (shear resistance) is developed. From the approximate boundary shown in Figure 7.3, it may be inferred that the asphalt specimen would not develop any deviator stress during construction if the asphalt core were constructed in a continuous manner with a strain rate lower than about 10^{-4} %/minute. If the deviator stress is zero, the lateral stress must be equal to the vertical stress like in a viscous fluid.

HAC is "supersaturated" with bitumen, and the aggregate skeleton structure in the asphalt concrete is weak. When the asphalt concrete is subjected to slow loading at a high temperature, the aggregate particles can rotate and slide inside the hot and viscous asphalt mastic and no deviatoric stresses are built up.

7.3 Stress-strain-strength relationships at various strain rates and temperatures

Zhang et al. (2021) performed unconfined compression tests in the strain rate range from 10^{2} second (60%/minute) to 10^{-5}/second (0.06%/minute) and in the temperature range from –30°C to 30°C. All the asphalt specimens (diameter of 100 mm) were drilled out of the asphalt core of the Gushan ACED in China during construction. The specimens were cut 100 mm high for the unconfined compression testing. Figures 7.4 and 7.5 show the test results for unconfined axial

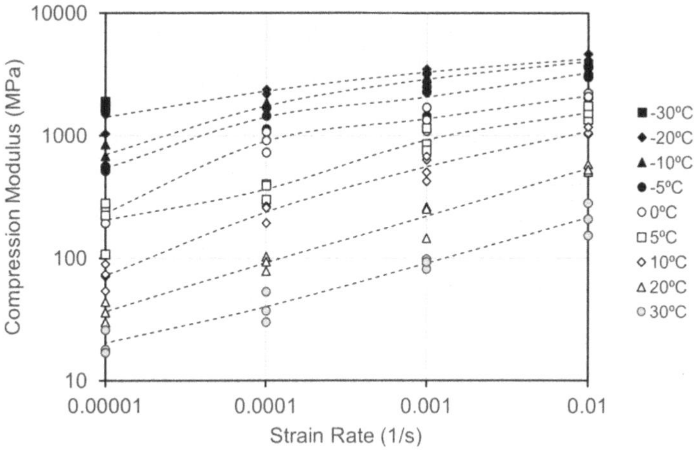

Figure 7.4 Unconfined compression modulus vs. compressive axial strain rate at various temperatures (Zhang et al., 2021).

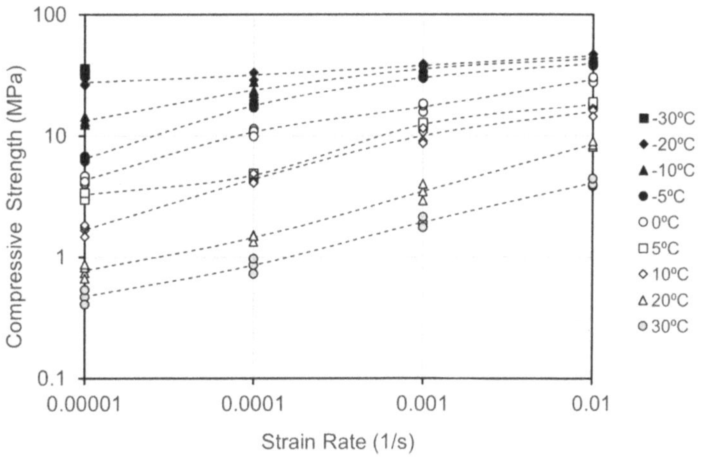

Figure 7.5 Unconfined compressive strength vs. compressive axial strain rate at various temperatures (Zhang et al., 2021).

compression modulus and unconfined axial strength. The secant modulus from 0% up to 1% axial strain was defined as the unconfined compression modulus, E_c. The figures show the values for E_c and unconfined compressive strength, σ_c, at various temperatures and loading strain rates.

Figures 7.4 and 7.5 show very consistent test results for unconfined compression modulus and unconfined compressive strength at various temperatures and strain rates. The modulus and strength are increased about 2–3 times at a temperature of –20°C and about 9–10 times at a temperature of 30°C when the strain rate is increased from 10^{-5} to 10^{-2}/second (1,000 times).

7.3.1 The TTS principle

The validity of the time-temperature superposition (TTS) principle is a basic assumption in the analysis of the effects of temperature and loading (strain) rate on the behaviour of asphalt concrete. The concept of a "master curve" allows the description of stress-strain-strength behaviour for a whole spectrum of temperatures and strain rates. The master curve is established by shifting separate modulus or strength curves along the strain rate axis. The relationship between these curves is described by the shift function. According to the TTS, if the relationship between modulus or strength and strain rate at a certain temperature is moved horizontally to a certain distance, it should follow the relationship at T_0, as is shown in Figure 7.6.

Thus, in a general notation:

$$E(\dot{\varepsilon},T) \text{ or } \sigma(\dot{\varepsilon}, T) = F(\log\dot{\varepsilon},T), E(\log\dot{\varepsilon},T_0) \text{ or } \sigma(\log\dot{\varepsilon},T_0) = F(\log\dot{\varepsilon},T_0) \tag{7.1}$$

where

E = modulus, MPa;
σ = strength, MPa;
$\dot{\varepsilon}$ = strain rate;
T = temperature, °C;
T_0 = standard (reference) temperature.
$F(T)$ is expressed as the translated (shifted) horizontal distance along the log $\dot{\varepsilon}$ axis, and then:

$$F(\log\dot{\varepsilon} + F(T),T_0) = F(\log\dot{\varepsilon},T) \tag{7.2}$$

Taking

$$F(T) = \log\varnothing(T) \tag{7.3}$$

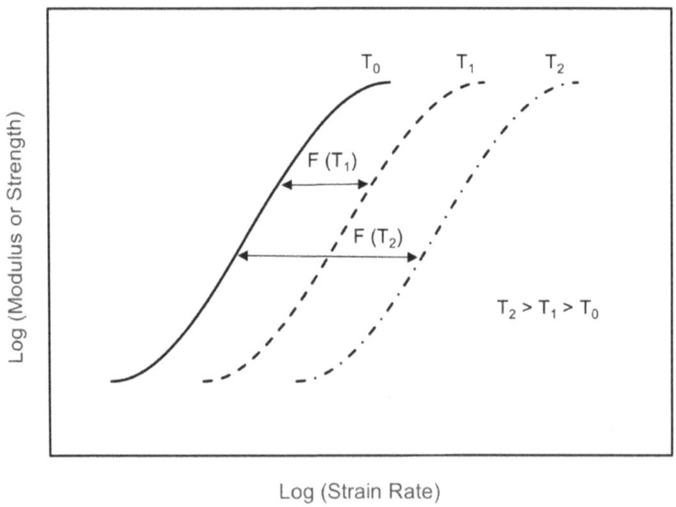

Figure 7.6 Schematic diagram of the time-temperature superposition (TTS) principle.

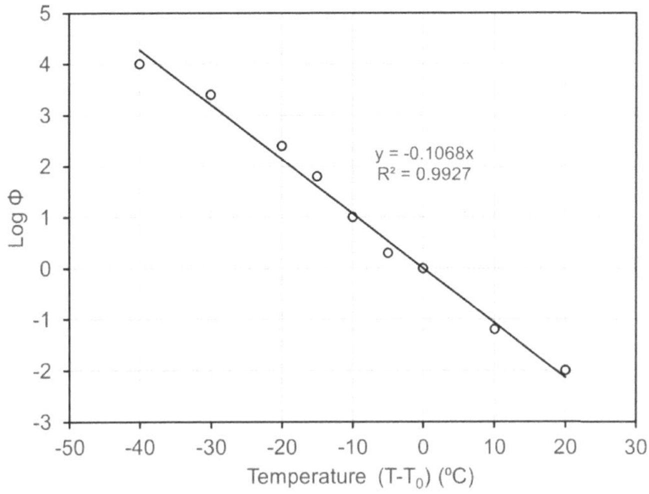

Figure 7.7 Relationship between the shifting function $\log \varnothing$ and temperature for unconfined compression modulus and compressive strength, T_0=10°C. In the equations shown in the figure, $y = \log \varnothing$ and $x = T - T_0$ (Zhang et al., 2021).

and combining Eq. (7.2) with Eq. (7.3), then:

$$F\left(\log\xi, T_0\right) = F\left(\log\dot{\varepsilon},\, T\right),\ \log\xi = \log\dot{\varepsilon} + \log\varnothing\left(T\right) \tag{7.4}$$

Equation (7.4) indicates that if $\dot{\varepsilon}$ is expressed as $\xi = \dot{\varepsilon}\,\varnothing\left(T\right)$, the E and σ curves at the standard temperature T_0 can be used to calculate the E and σ at temperature T. ξ is called the "reduced strain rate".

The specified reference temperature T_0 is taken as 10°C in the comparison of modulus and strength parameters. For unconfined compression tests, a regression analysis of the compression modulus and compressive strength test data shows that the shifting function $\log \varnothing$ vs. temperature gives a linear relationship, as shown in Figure 7.7 and by Eq. (7.5).

$$\log\ \varnothing\left(T\right) = -0.107\left(T - T_0\right) \tag{7.5}$$

When $T_0 = 10$°C,

$$\log\varnothing\left(T\right) = -0.107T + 1.07 \tag{7.6}$$

7.3.2 *Compression modulus and strength for other temperatures and strain rates*

Through the shift function Eq. (7.6), the master curves at 10°C are established by shifting the experimental data along the strain rate axis as shown in Figures 7.8 and 7.9. As the figures show nearly sinusoidal curves, the standard modulus and strength master curves were established by using four-term power-law series representations (M=4) and sine curve representations to

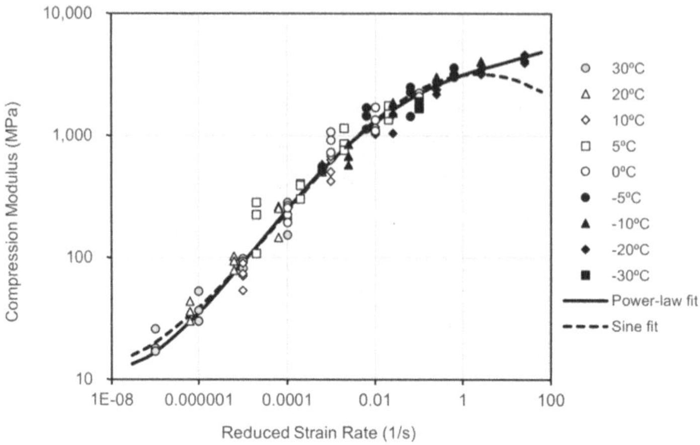

Figure 7.8 Compression modulus master curves at 10°C (Zhang et al., 2021).

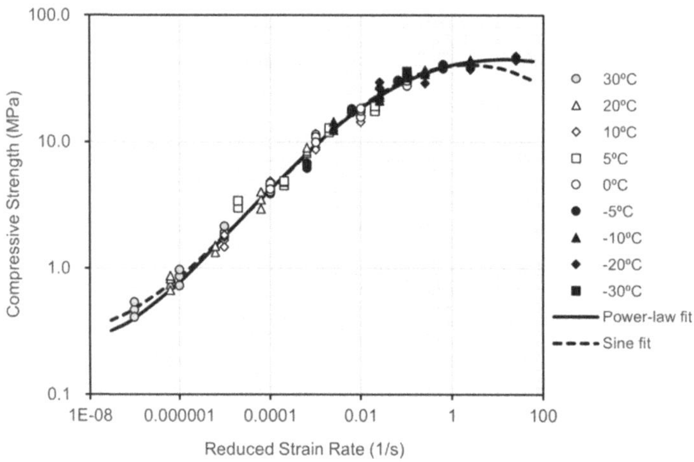

Figure 7.9 Compressive strength master curves at 10°C (Zhang et al., 2021).

smoothen the experimental data, as shown in Figures 7.8 and 7.9. Equations (7.7) and (7.8) show the modulus and strength master curves with the power-law fits.

$$\log E_c = 0.0009(\log\xi)^4 + 0.0068(\log\xi)^3 - 0.0252(\log\xi)^2 + 0.1236\log\xi + 3.4956 \quad R^2 = 0.987 \tag{7.7}$$

$$\log \sigma_c = 0.0005(\log\xi)^4 + 0.0017(\log\xi)^3 - 0.0415(\log\xi)^2 + 0.0837\log\xi + 1.6043 \quad R^2 = 0.991 \tag{7.8}$$

Thus, the compression modulus and compressive strength can be determined from Eqs. (7.7) and (7.8), respectively, for a given compressive strain rate and temperature.

Note: For the compression tests described in Section 7.3, the size of the specimens used to determine the unconfined compressive strength was 100 mm in diameter and only 100 mm in height (i.e., specimen height/diameter = 1). Thus, the compression modulus and compressive strength have been overestimated in the tests presented above because the specimens are shorter than recommended for compression specimens, and as the end caps were not frictionless, they provided some end restraints for the specimen to give increased compressive strength compared to a specimen with a height/diameter of 2, which is standard for such tests.

In a separate study, a group of specimens with a diameter of 100 mm were drilled out of the asphalt core of the Mile ACED in China during construction. They were cut into specimens of different heights 100, 150, 200, and 250 mm and tested at the same strain rate of 10^{-4}/second (0.6%/minute) and at the same temperature 10°C to investigate the effects of specimen height on the compression behaviour. The test results showed that the ratio of specimen height to diameter had no significant effect on the back-calculated unconfined compression modulus and the compressive strain at failure. However, when the ratio of specimen height to diameter was increased from 1 to 1.5, 2.0, and 2.5, the compressive strength was reduced by about 30% compared to the compressive strength of specimens with a ratio of 1.0.

In all the other triaxial tests reported herein, the specimen height has been 200 mm and the diameter 100 mm, i.e., height/diameter = 2.0.

7.3.3 *Tension modulus and tensile strength*

The effects of strain rate and temperature on the stress-strain-strength behaviour of HAC in tension were also investigated (Zhang et al., 2021). The tests were carried out in the strain rate range from 10^{-2} to 10^{-5} /second and in the temperature range from –30°C to 20°C.

The trends of the results for the tension modulus and tensile strength are like those for the compression behaviour. However, the shift function of the TTS has a coefficient of –0.23 for the tension tests, while it is about –0.11 for the unconfined compression tests (Figure 7.7). That indicates that the tension behaviour of asphalt concrete is significantly more sensitive to temperature changes than the compression behaviour.

As for the compression tests, the tension modulus and tensile strength master curves at 10°C were established by using three-term power-law series representations (M=2) to pre-smoothen the experimental data, as shown in Figures 7.10 and 7.11, respectively.

For the tension modulus,

$$\log E_t = -0.012(\log \xi)^2 + 0.21(\log \xi) + 3.353 \quad R^2 = 0.915 \tag{7.9}$$

For the tensile strength,

$$\log \sigma_t = -0.023(\log \xi)^2 + 0.07(\log \xi) + 0.783 \quad R^2 = 0.873 \tag{7.10}$$

Figure 7.10 shows that the tension modulus is continually increased when the tensile strain rate is increased from 10^{-6} to 10^{6}/second at 10°C. Figure 7.11 shows that the tensile strength is increased when the tensile strain rate is increased from 10^{-6}/second to about 0.1/second at 10°C. However, when the reduced tensile strain rate is larger than 0.1/second and the temperature is equal to or lower than –10°C, the tensile strain rate has no significant effects on the tensile

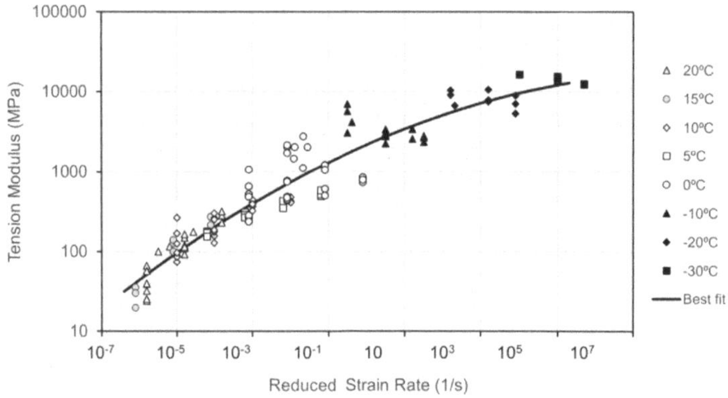

Figure 7.10 Tension modulus master curve at 10°C (Zhang et al., 2021).

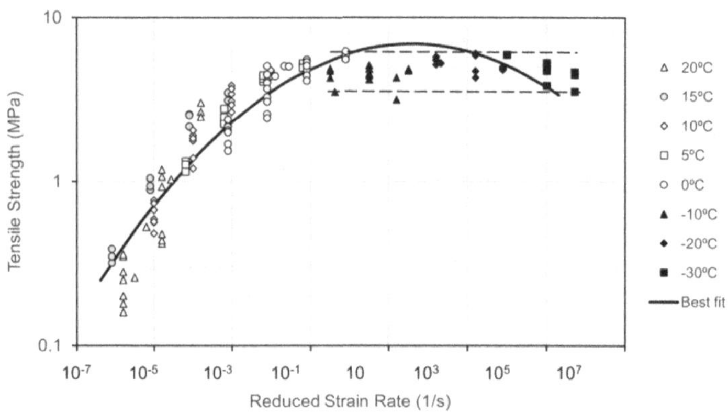

Figure 7.11 Tensile strength master curve at 10°C (Zhang et al. 2021).

strength which is between 3.5 and 6.0 MPa (between the dotted lines in the figure) with an average value of about 5.0 MPa.

Thus, the tension modulus and tensile strength can be determined from Eqs. (7.9) and (7.10) for a given tensile strain rate and temperature.

7.3.4 Stress-strain-strength parameters at various temperatures and strain rates

Based on the compression and tension test results at a few different temperatures and strain rates, shear strength parameters of HAC at other temperatures and loading strain rates can be estimated by combining the strength values obtained by the compression and tension tests (Zhang et al., 2021).

Typical stress-strain-strength properties and parameters for HAC at other temperatures and strain rates were calculated as shown in Table 7.1. The unit of the strain rate used in Table 7.1 is 1/second. When unit of 1/minute is used, the conversion is 1/minute = 1/60 seconds = 1.67%/

Table 7.1 Stress-strain-strength properties and parameters for hydraulic asphalt concrete (HAC) at various temperatures and strain rates (Zhang et al., 2021)

Temp. (°C)	*Parameter*	*Strain rate (1/second)*			
		10^{-5}	*10^{-4}*	*10^{-3}*	*10^{-2}*
30	Compression modulus (MPa)	16	31	79	216
	Compressive strength (MPa)	0.3	0.7	1.1	2.6
	Tension modulus (MPa)	2	5	12	30
	Tensile strength (MPa)	≈0	≈0	0.1	0.2
	Cohesion intercept (MPa)	N/A	N/A	0.2	0.5
	Friction angle (°)	N/A	N/A	56	59
20	Compression modulus (MPa)	33	85	231	578
	Compressive strength (MPa)	0.5	1.2	2.8	6.2
	Tension modulus (MPa)	16	37	82	174
	Tensile strength (MPa)	0.1	0.3	0.6	1.2
	Cohesion intercept (MPa)	0.2	0.4	0.9	1.9
	Friction angle (°)	42	37	40	43
10	Compression modulus (MPa)	91	247	613	1,281
	Compressive strength (MPa)	1.3	2.9	6.6	12.9
	Tension modulus (MPa)	101	209	412	767
	Tensile strength (MPa)	0.7	1.4	2.3	3.6
	Cohesion intercept (MPa)	0.7	1.4	2.7	4.8
	Friction angle (°)	17	20	29	34
0	Compression modulus (MPa)	265	649	1,339	2,260
	Compressive strength (MPa)	3.2	7.0	13.4	21.6
	Tension modulus (MPa)	488	895	1,554	2,554
	Tensile strength (MPa)	2.6	3.9	5.2	6.0[a]
	Cohesion intercept (MPa)	2.0	3.7	5.9	8.0
	Friction angle (°)	6	17	26	35
–10	Compression modulus (MPa)	687	1,398	2,327	3,254
	Compressive strength (MPa)	7.4	14.0	22.2	28.8
	Tension modulus (MPa)	1,780	2,882	4,417	6,406
	Tensile strength (MPa)	5.6	6.0[a]	6.0[a]	6.0[a]
	Cohesion intercept (MPa)	4.6	6.5	8.2	9.3
	Friction angle (°)	8	24	35	41
–20	Compression modulus (MPa)	1,458	2,395	3,315	4,190
	Compressive strength (MPa)	14.5	22.8	29.2	31.2
	Tension modulus (MPa)	4,895	6,997	9,464	12,111
	Tensile strength (MPa)	5.0[a]	5.0[a]	5.0[a]	5.0[a]
	Cohesion intercept (MPa)	6.0	7.5	8.5	8.8
	Friction angle (°)	29	40	45	46
–30	Compression modulus (MPa)	2,462	3,376	4,258	5,641
	Compressive strength (MPa)	23.3	29.5	31.2	28.7
	Tension modulus (MPa)	10,150	12,802	15,281	17,258
	Tensile strength (MPa)	5.0[a]	5.0[a]	5.0[a]	5.0[a]
	Cohesion intercept (MPa)	7.6	8.6	8.8	8.5
	Friction angle (°)	40	45	46	45

[a] Value obtained from actual test results.

second. It should be noted that the test results presented here for compression and tension tests are for unconfined stress conditions for the specimens.

It should be noted that the values in the table are valid for the specific asphalt concrete mix that had aggregates and filler of limestone, maximum grain size of aggregates 19 mm, filler content 13%, and the index of aggregate gradation was 0.41 (refer to Eq. 3.1). The bitumen content (B70) was 6.5%.

The stress-strain-strength parameters for other HAC mixes can be estimated by referring to the table at specific temperatures and strain rates in the preliminary design stage. However, actual parameters must be obtained by conducting tests for the specific asphalt concrete mix to be used in the project. The validity of the TTS principle was confirmed by the results of the extensive experimental test series by Feng et al. (2020c), Wang et al. (2020b), and Zhang et al. (2021) on the HAC mixes investigated at Xi'an University of Technology.

7.4 Long-term creep behaviour

The asphalt core in embankment dams is subjected to stepwise loading during construction, and step-by-step creep loading tests are considered appropriate to study creep behaviour of the asphalt core.

7.4.1 *Re-analysis of published creep test results*

When the first ACEDs were designed in the early 1960s, the main concern seems to have been the potential cracking of the core if the lateral stress support for the core from Zone 2 should decrease with time. Therefore, long-term creep triaxial tests were run to simulate that situation. The asphalt specimens were tested under constant vertical axial stress with decreasing radial stress. The mode of deformation of the triaxial specimen was axial compression with increasing deviator stress ($\sigma_1 - \sigma_3$), but the mean normal stress in the specimen was decreasing. The tests were run as multi-stage tests. The radial strain was allowed to increase in stages by decreasing the radial confining stress, and the resulting axial and radial strains were measured over a long period of time until a stable state of equilibrium was reached at that radial stress level (i.e., creep-stable state). Then, another decrement of radial stress was allowed.

Such tests were run and reported in connection with several ACED projects, e.g., Eberlaste Dam (Rienossl, 1973), Finsteral Dam (Schober, 1973; Pircher and Schwab, 1988), Megget Dam (Gallacher, 1988), and the more general studies by Haas (1983).

Wang (2008) reviewed and re-analysed the results of these early long-term creep tests. Figure 7.12 shows the results of the re-analysis of the experimental observations presented by Gallacher (1988) who ran creep tests with three constant axial stress levels of 0.8, 1.2, and 1.6 MPa, respectively. The relationship between deviator stress and axial strain at "creep-stable state" was found to be nearly linear for the imposed levels of deviator stress.

A summary of results from the back-calculation of "creep-stable modulus" (E_{cs}) from the experimental data published in the literature is shown in Table 7.2. E_{cs} is defined as the slope of the line in the stress-strain diagram in Figure 7.12 where the vertical axis shows the increase in deviator stress as the radial stress on the specimen is decreased. All the points shown are for situations at "creep-stable" states.

Table 7.2 shows that the back-calculated "creep-stable modulus" is in the range of 12–43 MPa for the various asphalt mixes and test conditions. The range in E_{cs} values is large. That may partly be attributed to the different compaction methods and different aggregate characteristics for the various specimens as discussed in Section 4.5.

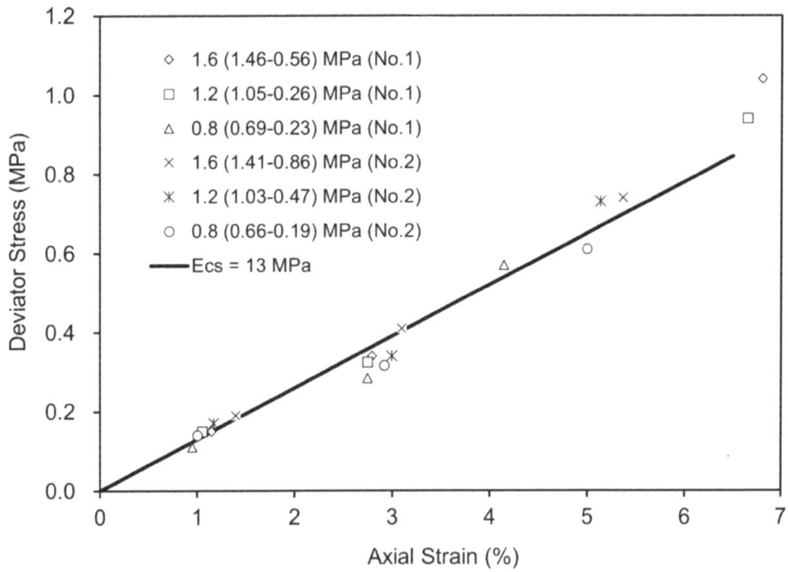

Figure 7.12 Deviator stress vs. axial strain at "creep-stable state". Data re-analysed from the long-term creep test results by Gallacher (1988). The stresses in the brackets give the decrease in radial confining stress during each test. The lateral stress decrease was made in small steps (Wang and Høeg, 2016).

Table 7.2 Values of "creep-stable modulus" (E_{cs}) calculated based on test results previously reported in the literature for various test conditions and asphalt concrete mixes (Wang and Høeg, 2016)

Reporter (first author)	*Reported year*	*Dam*	*Bitumen*		*Specimen size*		*Temp. (°C)*	*E_{cs} (MPa)*
			Grade	*Content (%)*	*Diameter (cm)*	*Height (cm)*		
Rienossl	1973	Eberlaste	B300	7.5	30	80	10	25
					12	30		43
Schober	1973	Finstertal	B65	6.1	30	80		39
Haas	1983		B65	6.1	12	30		12
Pircher	1988	Finstertal	B65	6.1–6.3	10	20		18
Gallacher	1988	Megget		6.8	15	36	10	13

Note: The specimens used by Pircher and Schwab (1988) were drilled out from the core in Finstertal Dam where the bitumen content was in the range of 6.1%–6.3%. The specimens in the other tests shown in the table were prepared and compacted in the laboratory.

7.4.2 Long-term creep tests at Xi'an University of Technology

Xi'an University of Technology has performed a series of long-term creep triaxial compression tests for asphalt concrete mixes for various ACEDs in China. Triaxial specimens with 100 mm in diameter and 200 mm in height were prepared either in the laboratory or were drilled out of the

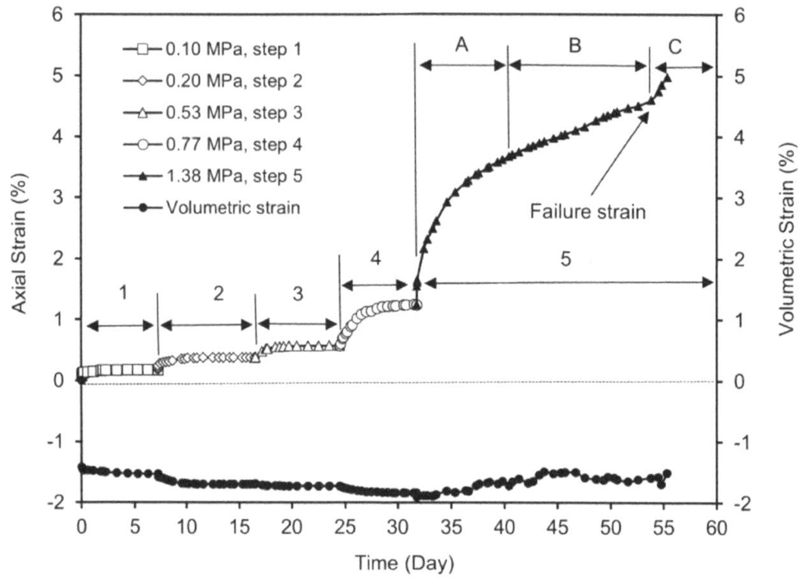

Figure 7.13 Axial strain and volumetric strain vs. time for a laboratory specimen under a constant radial confining stress of 0.5 MPa at 10°C and a stepwise increase in axial stress. Negative volumetric strain indicates reduction in volume.

compacted asphalt cores during dam construction. Initially, the axial stress (vertical stress) was equal to the radial stress on the triaxial specimens, i.e., no initial shear stresses in the specimens.

In each test, the axial stress was increased in steps while keeping a constant radial stress. For each level of imposed axial stress, the resulting axial strain and volumetric strain were measured until the axial strain reached a virtually constant value (i.e., creep-stable state) or it kept on increasing (i.e., creep-unstable state) towards failure under the last increment of axial stress. Typical results are shown in Figure 7.13. The two vertical axes show accumulated axial and volumetric strains, respectively, and the legend gives the magnitudes of the axial stress increments. The horizontal axis gives the duration (in days) of creep for each of the five (5) stress increments.

The figure indicates that the axial strain approached a constant value with time under low deviator stress (increments 1–4) but the axial strain kept on increasing under the large stress increment (number 5). When the large stress increment (number 5) was applied, the asphalt concrete responded by exhibiting a big increase in strain, and the stress-strain curve showed a complete creep process with three typical stages, A, B, and C. Stage A has a creep strain rate reducing with time, Stage B has a constant creep strain rate, and Stage C has a creep strain rate that increases with time. The inflection point where Stage B turns to Stage C defines the "failure strain". In the test shown in Figure 7.14, the failure strain was 4.7%.

The magnitude of compressive volumetric strains mainly depends on the compressibility of air voids in the asphalt concrete as the bitumen and mineral materials are nearly incompressible. The differences between air void contents of the various asphalt specimens are relatively large, and that causes the big differences in the volumetric strain when applying the first step of deviator stress. The first step in the deviator stress resulted in a big volumetric compressive strain of

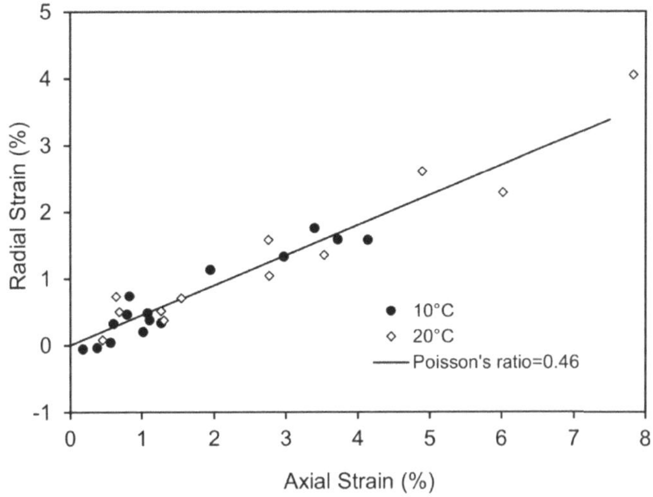

Figure 7.14 Radial strain vs. axial strain at "creep-stable state" in triaxial compression tests on the Kanerqi ACED asphalt mix.

1.4% for the specimen. After that, the volumetric strains responded with time in a similar way to the axial strain for the deviator stress increments 1–4 as shown in Figure 7.14. All the long-term creep triaxial test results at the creep-stable state indicated that the volumetric strains were negative (reduction in volume).

The results of the tests on the Kanerqi ACED asphalt specimens gave axial and volumetric strains during the tests, and the equivalent Poisson's ratio at the creep-stable state was found to be approximately 0.46 as shown in Figure 7.14.

Figure 7.15 shows axial strain "at failure" vs. radial confining stress level (see Figure 7.13). The relationship (Eq. 7.11) between the logarithm of failure strain and the confining stress is close to a straight line, and the figure indicates that the difference in testing temperature, 20°C vs. 10°C, has only little effect on the magnitude of recorded failure strains:

$$\log \varepsilon_f = a + b\frac{\sigma_3}{p_a} \tag{7.11}$$

where ε_f=axial failure strain (%); σ_3 =radial confining stress (MPa); p_a=atmospheric reference stress (0.1 MPa); and a, b=parameters for the given asphalt concrete mix. In this study, it was found that a=0.046 and b=0.424 with regression coefficient R=0.966.

The creep-stable modulus in the linear range is denoted as E_{cs}. A summary of results of creep-stable modulus from the long-term creep tests by the Xi'an University of Technology research team is shown in Table 7.3.

The "creep-stable modulus" (Table 7.3) is strongly influenced by the properties of the asphalt concrete mixes, the test temperature, the flakiness index (FI), and the compaction equipment and method. The Simulating Field Compaction (SFC) compaction method in the laboratory yields asphalt specimens with stress-strain behaviour much like that of the drilled core specimens in the field. As discussed in Section 4.3, field roller compaction yields asphalt concrete

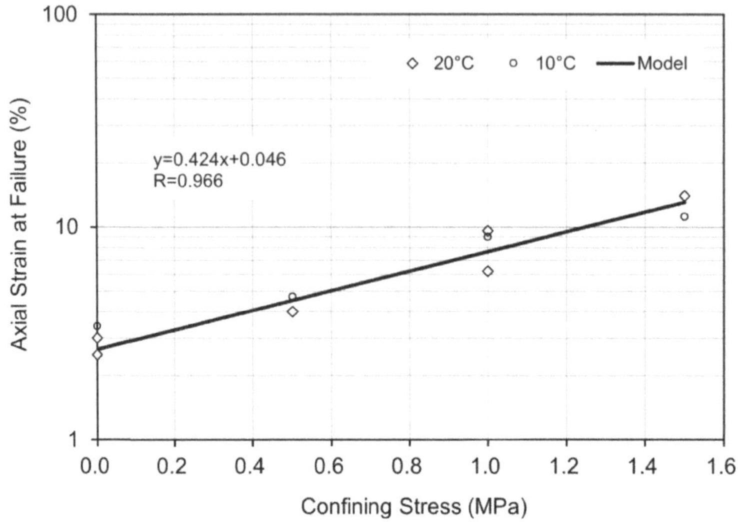

Figure 7.15 Logarithm of axial failure strain vs. confining stress for the Kanerqi asphalt specimens at 20°C and at 10°C. In the equation shown in the figure, $y = \log \varepsilon_f$ and $x = \sigma_3/p_a$.

Table 7.3 Values of "creep-stable modulus" (E_{cs}) from the tests by Xi'an University of Technology for various test conditions and asphalt concrete mixes

Reporter (first author)	*Reported year*	*Dam*	*Bitumen*		*Specimen preparation*	*Temp. (°C)*	E_{cs} *(MPa)*
			Grade	*Content (%)*			
Wang	2009	Kanerqi	B70	6.5	Laboratory[a]	5	57
				6.1		10	50
				6.1		20	25
				7.0		20	10.5
				7.8		20	6.5
Feng	2020	Laojiaoxi	B70	6.5	Field core	20	10
						30	4
Wang	2023	–	B70	7.0	Laboratory[b]	10	6.0
			B110			10	4.5
Wang	2024	Quxue	B70	6.4	Field core	5	15
						10	13
						15	10
						20	8
						25	6

[a] The specimens were prepared by using the Marshall compaction method for triaxial specimens using a Marshall hammer (see Section 4.1).

[b] The specimens were prepared by using the SFC compaction method (see Section 4.5).

with significantly lower compression modulus than laboratory-compacted triaxial specimens prepared in the standard Marshall cylindrical mould with an inner diameter of 101.6 mm.

Based on the results from the long-term creep triaxial tests, Wang (2024) proposed an asphalt concrete stress-strain-time-temperature creep model (SSTTC model), as described in Section 7.8.

7.5 Behaviour during slow unloading-reloading cycles

During dam operation, the asphalt core is subjected to unloading-reloading cycles due to raising and lowering of the reservoir. All triaxial specimens referred to in this part of the study were first subjected to long-term creep and had reached "creep-stable state" before the specimens were subjected to unloading-reloading cycles.

The axial unloading was done in steps, keeping the confining stress constant. When an axial unloading step was applied, the axial strain decreased, and after it reached a "creep-stable state" in a time usually less than 3 minutes, the next axial unloading step was applied. When the deviator stress was reduced by this unloading, the asphalt concrete behaved almost elastically as the aggregate skeleton structure played a predominant role in the unloading.

A similar test procedure was then followed during reloading back to the initial deviator stress keeping the confining stress constant. As for the axial unloading steps, it took less than 3 minutes to reach a stable strain state. The asphalt concrete behaved almost elastically. Typical results after three unloading-reloading cycles with a confining stress of 0.8 MPa and a test temperature of 10°C are shown in Figure 7.16.

Figure 7.16 shows that the three unloading-reloading cycles followed almost the same paths. There was no axial strain accumulation during the cycling, and there was little hysteresis. The average unloading-reloading modulus, E_{ur}, is defined as the slope of the secant from the lowest to the highest deviator stress applied during cycling.

The unloading-reloading modulus was in the range of 500–1,500 MPa for confining stresses in the range of 0.3–1.8 MPa at 10°C. E_{ur} may be expressed as:

$$E_{ur} = K_{ur} p_a \left(\frac{\sigma_3}{p_a} \right)^n \tag{7.12}$$

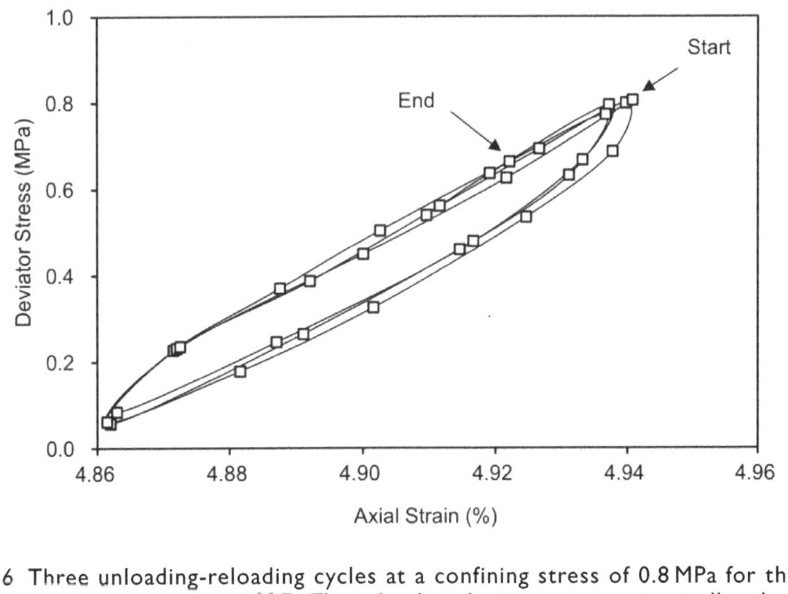

Figure 7.16 Three unloading-reloading cycles at a confining stress of 0.8 MPa for the Kanerqi asphalt specimen at 10°C. The reloading deviator stress was smaller than the previous maximum deviator stress at creep-stable state.

where E_{ur}=modulus under unloading-reloading condition (MPa); p_a=atmospheric reference pressure (0.1 MPa); σ_3=confining stress (MPa); and K_{ur} and n=test parameters obtained by regression analysis. For the asphalt concrete mix in this study, it was found that K_{ur}=5315 and n=0.298, with R=0.745.

Tests were also run at 20°C, and the recorded unloading-reloading modulus was found to vary between 430–520 MPa for confining stresses in the range of 0.5–1.5 MPa.

All the unloading-reloading cycles were run in such a way that the reloading deviator stress never exceeded the previous deviator stress applied in the creep-stable state. If the applied reloading stress had been increased beyond that level, further creep strains would have occurred. Typical unloading-reloading curves are shown in Figure 7.17. The unloading started from a creep-stable state, and the reloading also reached a creep-stable state.

A comparison of the unloading-reloading modulus (E_{ur}) and the long-term creep-stable modulus (E_{cs}) for the Kanerqi asphalt specimens tested at 10°C and 20°C is presented in Table 7.4.

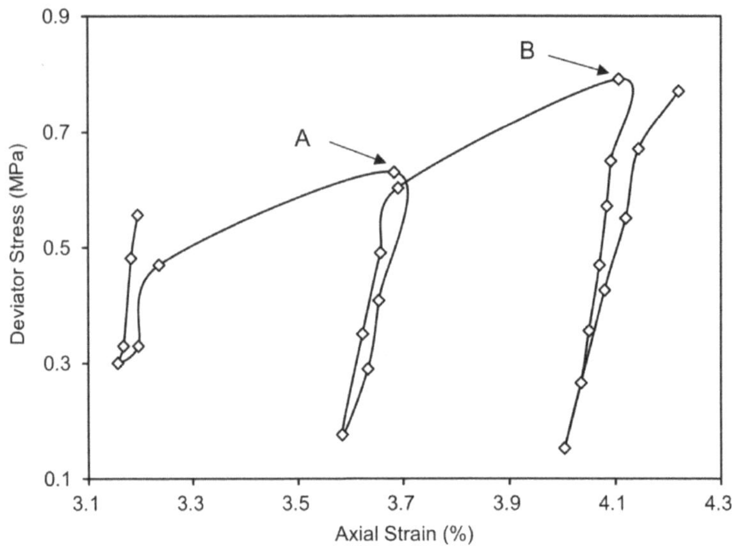

Figure 7.17 Typical unloading-reloading curves at creep-stable state for the Kanerqi ACED asphalt specimen at 20°C. The deviator stress at point A was larger than the previously applied deviator stress, and the deviator stress at point B was larger than the deviator stress applied in the preceding cycle.

Table 7.4 Comparisons of unloading-reloading modulus and long-term, creep-stable modulus for the Kanerqi asphalt specimens (Wang and Høeg, 2016)

Temperature (°C)	*Long-term creep-stable modulus E_{cs} (MPa)*	*Unloading-reloading modulus E_{ur} (MPa)*	*Ratio of E_{ur}/E_{cs}*
10	50	500–1,500 (Confining stress 0.3–1.8)	10–30
20	25	430–520 (Confining stress 0.5–1.5)	17–21

The unloading-reloading modulus was found to be about 20 times higher than the long-term creep-stable modulus. When asphalt specimens are subjected to long-term creep loading, the aggregates in the specimen are rearranging, and the edges and corners of aggregate particles are getting in contact with each other (Liu et al., 2022). With time, this forms a strong and stiff skeleton structure and achieves a creep-stable state of equilibrium. When applying unloading and reloading deviator stresses smaller than the previous maximum deviator stresses at creep-stable state for the unloading-reloading cycles, the stiff aggregate skeleton structure plays a dominant role in the behaviour with a very high modulus (Figures 7.16 and 7.17). When applying a deviator stress bigger than the previous maximum deviator stress, the strong and stiff aggregate skeleton structure that has been formed is redistributing and creates a "new" long-term creep strain with creep-stable modulus (points A and B in Figure 7.17).

For pump storage installations, the cyclic unloading-reloading may take place daily, but it is still classified as slow loading as opposed to dynamic cyclic loading caused by earthquakes.

7.6 Behaviour during dynamic cyclic loading

The behaviour of the asphalt core during and after dynamic cyclic loading must be investigated for ACEDs in seismic regions. Cyclic loading tests were carried out on three asphalt core mixes for dams in China (Wang and Høeg, 2011). The testing temperatures used were the local annual average temperatures at the three project sites: 3.5°C at Xiabandi Dam, 9°C at Qiapuqihai Dam, and 20°C at Longtoushi Dam.

Other researchers have tested other mixes (Feizi-Khankandi et al., 2008; Akhtarpour and Khodaji, 2013; Han et al., 2023; Meng et al., 2023).

Before the dynamic, cyclic loading was applied, radial confining stress and axial stress were imposed in small simultaneous steps on the specimen until the specified initial static stress state with the stress ratio of $K_c=\sigma_1/\sigma_3$ was reached. After the specified duration of the applied static state, the cyclic axial loading (sinusoidal) was applied with a frequency of 1 Hz. In each cyclic test, the confining stress (σ_3) was kept constant, and cyclic loading was only applied in the axial direction. The cyclic axial stress was increased in steps, and five stress cycles were applied at each level of cyclic stress. For the cyclic triaxial tests, the half-amplitude of the imposed cyclic stress and the corresponding half-amplitude of the cyclic strain are denoted by σ_d and ε_d, respectively. Under a small initial static deviator stress (e.g., $\sigma_3=500$ kPa and $K_c=1.2$) and a relatively small cyclic axial stress (e.g., $\sigma_d=270$ kPa), the resulting axial strain was virtually elastic, and the residual strain after the cyclic loading was negligible. Under a larger initial deviator stress (e.g., $\sigma_3=500$ kPa and $K_c=2.0$) and larger cyclic loading stress (e.g., $\sigma_d=550$ kPa), the resulting axial cyclic strain was still approximately constant, but there was also some residual strain. The residual strain was mainly caused by the sustained deviator stress (Wang and Høeg, 2011, 2016).

For the conventional triaxial cyclic test equipment, the available maximum cyclic loading stress σ_d was always specified to be smaller than the specified sustained axial stress σ_1, i.e., one-way cyclic axial loading. Two-way cyclic shearing with stress reversals is difficult to achieve with conventional triaxial test equipment but may be performed in cyclic simple shear tests and in torsional cyclic shear tests as described in Section 7.7.

Figure 7.18 shows a typical cyclic stress-strain hysteresis loop from a triaxial test with one-way cyclic loading.

The cyclic modulus (E_d) from the triaxial compression test is calculated as:

$$E_d = \sigma_d / \varepsilon_d \tag{7.13}$$

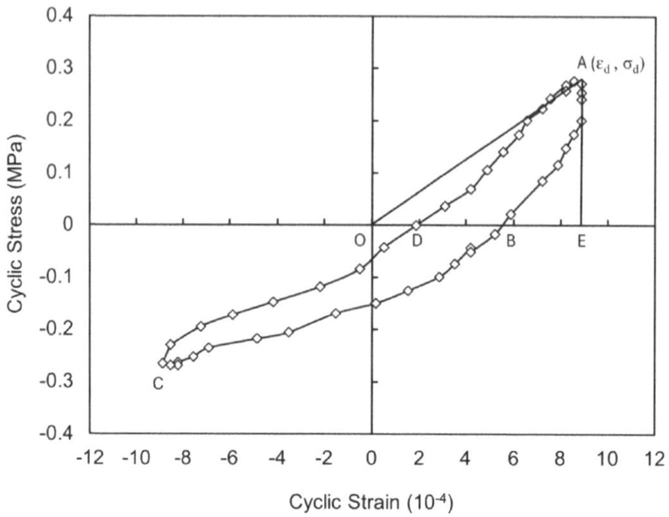

Figure 7.18 Typical cyclic stress-strain hysteresis loop.

The damping ratio (λ_d) is calculated as:

$$\lambda_d = \frac{1}{4\pi}\frac{S}{S_\Delta} \tag{7.14}$$

where S=area enclosed by the hysteresis loop in a single loading cycle, and S_Δ=area of the triangle OAE in Figure 7.18.

The shape and area of the cyclic stress-strain hysteresis loop depends on the level of sustained static stress state, temperature, and magnitude of imposed cyclic stress on a given asphalt concrete mix. For a low static stress state and at high temperature, the cyclic modulus on the extension side is somewhat lower than that on the compression side. Figure 7.18 shows such an example in which the static confining stress was 500 kPa, and the stress ratio was 1.2 at 20°C. For a high static stress state and at low temperature the extension cyclic modulus and compression cyclic modulus are nearly identical. This agrees with the findings of Feizi-Khankandi et al. (2008). In the following presentation and discussion of results, the cyclic modulus (E_d) is taken as the average value of extension and compression cyclic moduli determined from the cyclic stress-strain hysteresis loops.

Typical imposed cyclic stress (σ_d) vs. resulting cyclic strain (ε_d) for the Longtoushi asphalt concrete material at different sustained static stress states and at 20°C is shown in Figure 7.19a. The damping ratio computed from the hysteresis curves is shown in Figure 7.19b. The figure shows that, for all the static and cyclic stress combinations, the cyclic stress σ_d vs. cyclic strain ε_d gives an almost linear relationship. It is only when the imposed σ_d is approaching the sustained static stress σ_1 that the σ_d vs. ε_d diagrams show a tendency to deviate from a straight line. Under all the different stress states and temperature conditions tested, the maximum cyclic strain was less than 0.3%, even when the high static deviator stress and cyclic stress levels were applied.

Figure 7.19 shows that the computed damping ratios were between 0.1 and 0.25 and a reduction in damping ratio with increasing cyclic strain. On the contrary, some experimental results by other investigators on other asphalt mixes show that the damping ratio increased somewhat

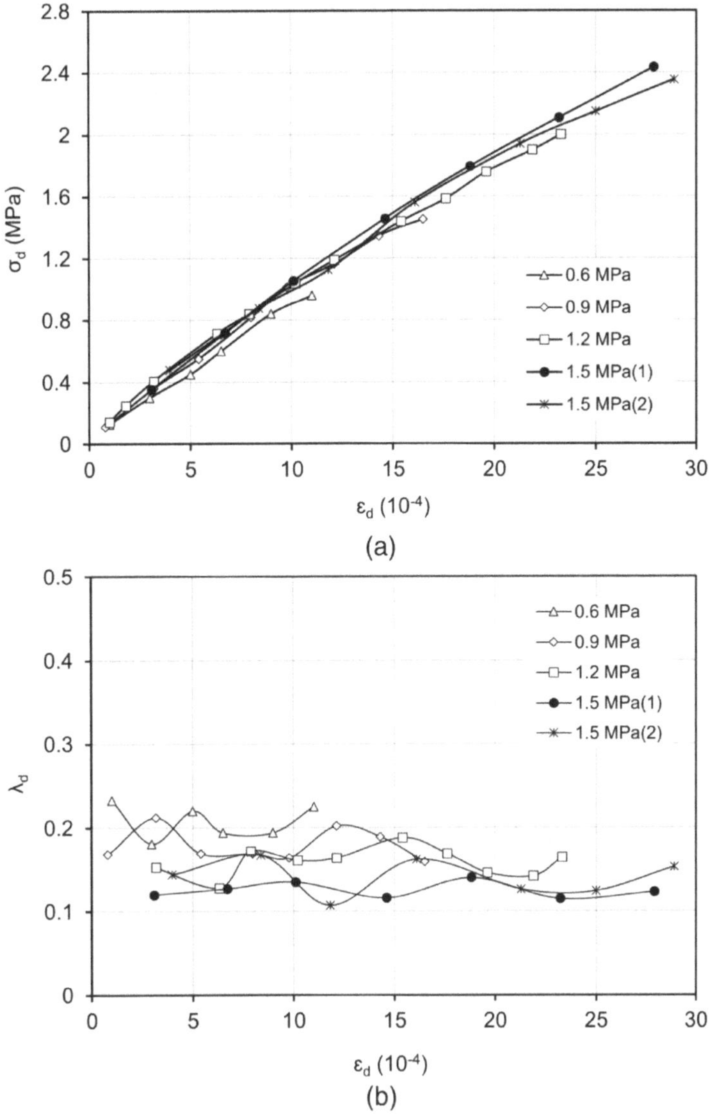

Figure 7.19 Cyclic stress (a) and damping ratio (b) vs. cyclic strain for $K_c = 1.8$ and for various values of confining stress for the Longtoushi ACED asphalt specimens at 20°C.

with cyclic strain level. Considering all the tests performed and reviewed in the study, the damping ratio was found to lie between 0.07 and 0.30, depending on the stress state and the temperature level. This range of values agrees well with that obtained by Feizi-Khankandi et al. (2008) and Han et al. (2023).

For the Xiabandi ACED asphalt specimens tested at 3.5°C, the cyclic loading modulus was found to be very high, between 2,000 and 2,800 MPa. For the Qiapuqihai ACED specimens at 9°C, the modulus was found to be between 1,100 and 2,400 MPa, and for the Qiapuqihai and Longtoushi ACED specimens at 20°C, the modulus was found to be between 730 and 1,100 MPa.

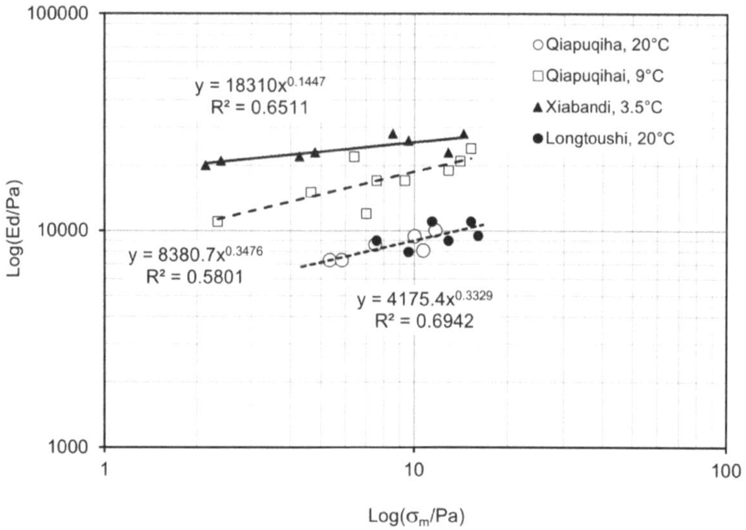

Figure 7.20 Log (E_d/p_a) vs. log (σ_m/p_a) for three asphalt core materials at 3.5°C, 9°C, and 20°C, respectively. In the equations shown in the figure, $y = E_d/p_a$ and $x = \sigma_m/p_a$.

When the cyclic modulus E_d is plotted vs. mean stress σ_m in a logarithmic diagram, as shown in Figure 7.20, the test results show an almost linear relationship. The expression is:

$$E_d = kp_a(\sigma_m/p_a)^n \tag{7.15}$$

where:

σ_m = mean sustained stress, $\sigma_m = (\sigma_1 + 2\sigma_3)/3$, MPa,
p_a = atmospheric pressure, 0.1 MPa (a reference stress level),
k, n = test parameters determined by regression analysis of the experimental data (see values in Figure 7.20).

Figure 7.20 shows that temperature had a very significant effect on the cyclic modulus. At a mean sustained stress of 1.0 MPa, the cyclic loading modulus at 20°C was about 900 MPa, at 9°C, about 1,900 MPa, and at 3.5°C, about 2,500 MPa. Figure 7.20 also shows that the cyclic moduli at 20°C for the Qiapuqihai and Longtoushi asphalt concrete specimens agreed quite well with each other.Akhtarpour and Khodaii (2013) investigated the cyclic shear modulus of asphalt concrete under a confining stress of 0.5 MPa at 22°C and found that the cyclic shear modulus was between 150 and 250 MPa for bitumen contents between 5.5% and 7.0% in the cyclic strain range between 0.1% and 0.3%.

Han et al. (2023) investigated the cyclic modulus of asphalt concrete under confining stresses in the range of 0.5–0.8 MPa, stress ratios in the range of 1–1.6 at temperatures in the range of 4°C–22°C. The bitumen content (B70) was 6.1%. Their test results agree well with the results that are shown in Figure 7.20.

During the cyclic triaxial tests, the specimen volume change was measured, and Poisson's ratio could be computed. The computed average Poisson's ratio during cyclic loading was found to be 0.35 considering all the specimens (Wang and Høeg, 2011).

In the tests described in the preceding sections, five (5) load cycles were applied at each cyclic stress level. Only very small residual strains were recorded after the five cycles of loading. In a supplementary series of tests, many cycles were imposed on Qiapuqihai ACED specimens at various stress states and at 9°C. Prior to cyclic loading, the specimens were first allowed to creep for 40 min under the sustained static stress condition. Significant creep occurred, but the period was not long enough to reach a fully "creep-stable state," as discussed in Section 7.4. Then, the cyclic loading was imposed, and the elastic and residual strains were measured. During more than 6,000 cycles of loading, the cyclic strain magnitude stayed constant. Even for this high cyclic stress of 1.0 MPa, the number of cycles had virtually no effect on the magnitude of cyclic strain. However, the residual axial strain increased with the number of cycles. The residual strain recorded was predominantly creep strain caused by the sustained deviator stress. Feizi-Khankandi et al. (2008) and Akhtarpour and Khodaii (2013) ran up to 10,000 cycles of loading and found that there was no visible damage (fissures or cracks) observed in the specimens.

To compare the post-cyclic behaviour of specimens subjected to cyclic loads to that of specimens that had not been subjected to cyclic loads, a special test series was run.

Three groups of specimens were first subjected to the same static sustained stress state with confining stress 1.0 MPa and axial stress 1.7 MPa for 5 hours. Monotonic loading tests (unconfined) were carried out on one of the three groups of specimens. The permeability coefficients and the results from the monotonic tests were later taken as reference values for comparison. Two groups of specimens were subjected to cyclic loading. The cyclic axial stress σ_d was increased, in ten steps, up to 1.5 MPa, and at each stress level, 30 load cycles with a frequency of 1 Hz were applied. The two groups of specimens were dismounted from the triaxial cell after being subjected to cyclic loading. One of the two groups of specimens was used for permeability tests and another for monotonic loading tests. The test results indicate that the cyclic loading has had virtually no degradation effect on the stress-strain-strength behaviour of the asphalt concrete. The permeability test results also show that the permeability coefficient of the specimens subjected to the static and cyclic loading did not increase. Thus, no fissures or cracks could have opened (see Section 3.5).

Meng et al. (2023) investigated the cyclic behaviour of HAC in tension using loading frequencies 0.5, 1, 2, and 5 Hz at 5°C. The test results indicate that the difference in frequencies had insignificant effects on the tensile cyclic behaviour. The tensile cyclic loading had virtually no degradation effect on the tensile stress-strain-strength behaviour although a large creep tensile strain was reached under the sustained loading.

Zhao et al. (2012) investigated the triaxial cyclic modulus of asphalt mixes for road pavements using seven (7) different loading frequencies from 0.1 to 25 Hz at five (5) different temperatures from 4°C to 55°C. The test results indicate that the cyclic modulus increases significantly with increasing frequency.

7.7 Torsional shear testing of asphalt concrete

In a conventional cyclic triaxial test apparatus, when the cyclic stress σ_d is more than the specified sustained, static axial stress σ_1, the resulting axial stress will be in tension, and the head in the triaxial test cell will disjoint from the top of the specimen. Therefore, for such triaxial tests, the available maximum cyclic loading stress σ_d is always specified to be smaller than the specified sustained axial stress σ_1. Under such cyclic stress levels, the cyclic stress-strain curves are unable to reach the cyclic peak strength as only one-way cyclic loading can be achieved.

Feng (2018) used a torsional triaxial shear test apparatus, often applied in geotechnical engineering, to investigate the cyclic shear stress-strain-strength behaviour of HAC. The asphalt specimen was a hollow cylinder. The specimen was prepared with dimensions 100 mm in outer diameter and 200 mm in height in a split steel mould. After the specimen in the mould was cooled down to room temperature, it stayed at 5°C for 24 hours. The specimen was kept constrained in the mould while an axial core of 54 mm in diameter was drilled out through the axis of the specimen, cutting through the aggregates without noticeable disturbances of the asphalt specimen.

The hollow cylindrical asphalt specimens were subjected to reciprocating (reverse direction) horizontal torsional shear stress under specified vertical axial stress and confining stress at 20°C. The magnitude of the torsional shear stress with a frequency of 1 Hz was increased in steps until the specimen fractured/failed. Torsional shear specimens mostly failed near the top end of the grips with a roughly horizontal shear fracture surface due to the unavoidable stress concentrations at the ends of the grips.

The confining stresses applied were 0.2, 0.4, 0.6, and 0.8 MPa, while the stress ratios (vertical stress/confining stress) were 1.0, 1.2, 1.4, and 1.6. The test results indicated an approximately linear relationship between torsional shear stress and shear strain. The torsional shear modulus was found to be between 80 and 120 MPa depending on the applied test conditions. The shear strength and shear strain at failure were about 3.0 MPa and about 2.5%, respectively (Feng, 2018).

Based on the experience from these introductory tests, the authors conclude that it is entirely feasible to prepare virtually undisturbed hollow-cylinder specimens of HAC and perform static and dynamic cyclic torsional shear stress – shear strain tests. Torsional shear tests may be used in future research investigations of static and dynamic asphalt behaviour including two-way cyclic loading conditions that cannot be achieved in the conventional triaxial apparatus.

7.8 Proposed creep model for HAC behaviour in dam core

7.8.1 Introduction to constitutive model

The loading of the asphalt core is described in Section 7.1. The stresses and strains in the core, both magnitude and distribution, should be analysed to investigate whether there are zones subject to cracking and possibly local failure. Special attention must be given to interfaces with appurtenant structures (e.g., spillways and gates) and irregular foundation and abutment profiles where stress and strain concentrations may occur. HAC tends to relieve itself of local stress and strain concentrations due to its viscoelastic and ductile behaviour, but the overstressing cannot be accommodated if the overstressed area becomes too large and intense for the neighbouring areas to "absorb" the additional stresses (see Section 5.4). The dam and core design and the asphalt concrete mix must be modified if tolerable limits are exceeded, and critical cracking may occur.

Long-term triaxial creep tests may be run on asphalt concrete specimens in the laboratory to simulate the gradual construction of the asphalt core in the field and the stress conditions in different parts of the core. Typical stress-strain behaviour as function of temperature and loading rate as well as values for strength and tolerable strain before cracking occurs may be defined by performing tests on the asphalt concrete mix proposed for the project (see Chapter 3).

Test results show that the creep strains caused by incremental stresses below a yield boundary surface approach a constant strain level which is denoted as the "creep-stable state". The relationship between imposed deviator stress and axial strain at "creep-stable state" is found to be nearly linear, and the linear relationship at creep-stable state is defined by the creep-stable modulus, E_{sc}. Tables 7.2 and 7.3 show creep-stable modulus values for various test conditions and various asphalt concrete mixes. The creep-stable modulus may be treated as an equivalent elastic modulus in the linear creep-stable model proposed by Wang and Høeg (2016). The model may be used together with a numerical stress-deformation analysis procedure to estimate the stresses and strains in various parts of the asphalt core.

Wang (2024) performed long-term triaxial creep tests on asphalt specimens drilled out of the cores of Quxue ACED and Laojiaoxi ACED. The tests were run at various radial confining stress levels in the range of 0.5–1.5 MPa and at various temperatures in the range of 5°C–30°C. The relationship between the back-calculated creep-stable modulus and temperature showed a nearly linear relationship in the temperature range of 5°C–30°C (Figure 7.21). The trend of the relationship seems clear from the figure, but there is scatter in the results from the average value at any specific temperature, and more test results are needed to investigate the assumption of linearity.

Test results show that there is a stress state boundary where the asphalt concrete changes from creep-stable to creep-unstable behaviour. The boundary is defined as the "plastic yield boundary", named PYB by Wang and Høeg (2016), and can be approximated by using the results from long-term triaxial creep tests. Applying the Mohr-Coulomb type strength criterion, Wang (2024) computed the equivalent "cohesion intercept" and "angle of shearing resistance", vs. temperature for the PYB as shown in Figure 7.22. PYB is then serving a function like that of the Mohr-Coulomb failure envelope in soil mechanics.

More test results are required to investigate the assumption of linearity (Figure 7.21) and the magnitudes of "angle of shearing resistance" and "cohesion" as shown in Figure 7.22.

The asphalt core in the centre region of the embankment dam is placed and compacted in a hot state. The lateral loads are increasing on the asphalt core when raising the embankment and filling the reservoir. Figure 7.21 shows that the creep-stable modulus E_{cs} is less than about 10 MPa when the temperature in the asphalt core is higher than about 15°C. The behaviour of the warm asphalt core at this stage is mainly governed by the properties of the asphalt mastic as the aggregate skeleton structure, which gradually will be formed by the coarse aggregates when the temperature cools down, is still weak. At this temperature, the angle of shearing resistance of

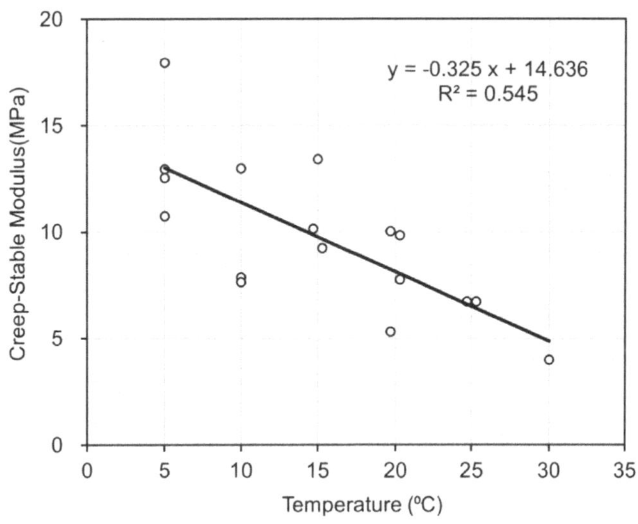

Figure 7.21 Back-calculated creep-stable modulus E_{cs} vs. temperature for specimens drilled out of the Quxue and Laojiaoxi ACED cores.

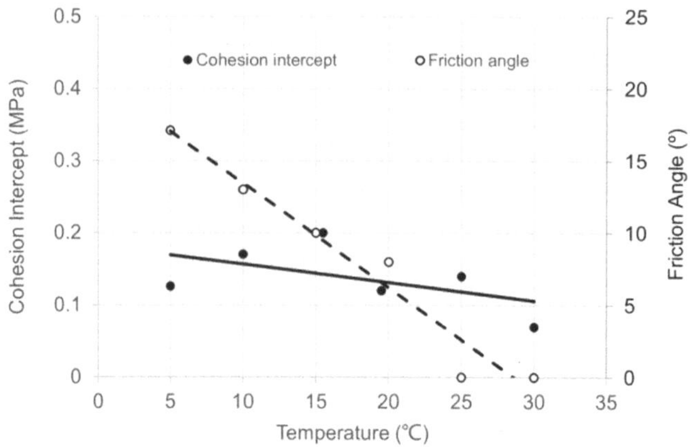

Figure 7.22 "Cohesion intercept" and "angle of shearing resistance" vs. temperature.

the asphalt concrete is low, and the creep-stable modulus E_{cs} is low. The temperature in the asphalt core reduces with time, and the asphalt mastic stiffness and bonding between aggregates increase. The angle of shearing resistance of asphalt concrete depends on the interlocking of particles in the aggregate skeleton structure and on the friction between aggregate surfaces. When the temperature in the asphalt concrete decreases, the bonding between the bitumen and aggregates increases, and the aggregate skeleton structure is strengthened.

The creep-stable modulus E_{cs} of the asphalt core just after compaction is about 1/10 that of the adjacent transition zones (Zone 2) but increases with cooling. As long as the asphalt concrete is hot, the likelihood of cracking is low and the ability to self-heal high. The core strength and stiffness are low, and the core depends on the stability and the support of Zone 2.

7.8.2 Mathematical formulation of stress-strain-time-temperature creep model

As shown in Figure 7.21, the relationship between the creep-stable modulus E_{cs} and temperature seems to be approximately linear, and as a first approximation, the linear viscoelastic theory is used to establish the asphalt concrete stress-strain-time-temperature creep (SSTTC) model, as presented by Wang (2024).

The long-term triaxial creep test results on the Quxue and Laojiaoxi ACED core specimens indicate that the axial strain approaches a constant value with time (i.e., creep-stable state) for low deviator stresses, and the stress-strain-time curves can be modelled with the generalized Kelvin model of the viscoelastic theory (Tong et al., 2022; Wang, 2024). The axial strain under high deviator stress increases at an increasing rate until failure (i.e., creep-unstable state), and the stress-strain-time curves may be modelled using the generalized Maxwell or Burgers models of the viscoelastic theory. As a first stage in the development, the linear form of the creep compliance for the generalized Kelvin model is expressed as a Prony Series:

$$D(t) = D_g + \sum_{i=1}^{n} D_i \left(1 - e^{-\frac{t}{\tau_i}}\right) \tag{7.16}$$

where D_g=the instantaneous compliance; D_i and τ_i=retardation strengths and retardation times, respectively, of the ith Kelvin element in the generalized model; and t=time. In the generalized Kelvin model, only the independent spring element (D_0) carries the stress after an infinite creep time (t=∞). Therefore, the long-term compliance D_0 can be expressed as:

$$D_0 = D_g + \sum_{i=1}^{n} D_i \tag{7.17}$$

For the long-term triaxial creep tests, the creep axial strain $\varepsilon(t)$ can be expressed as:

$$\varepsilon(t) = (\sigma_1 - \sigma_3) D(t) \tag{7.18}$$

Where $\varepsilon(t)$=axial creep strain, %; $\sigma_1 - \sigma_3$=deviator stress, MPa.

Equation (7.16) can be written as:

$$D(t) = \frac{1}{E_g} + \sum_{i=1}^{n} \frac{1}{E_i}\left(1 - e^{-\frac{t}{\tau_i}}\right) \tag{7.19}$$

where E_g=the instantaneous elastic modulus and E_i=retardation modulus. The long-term compliance D_0 can be expressed as:

$$D_0 = \frac{1}{E_{cs}} = \frac{1}{E_g} + \sum_{i=1}^{n} \frac{1}{E_i} \tag{7.20}$$

E_{cs} is defined as the long-term creep-stable modulus by Wang and Høeg (2016).

When the viscoelastic parameters are expressed as the functions of temperature, the equations are expressed as:

$$\varepsilon(t, T) = (\sigma_1 - \sigma_3) D(t, T) \tag{7.21}$$

$$D(t, T) = \frac{1}{E(T)_g} + \sum_{i=1}^{n} \frac{1}{E(T)_i}\left(1 - e^{-\frac{t}{\tau(T)_i}}\right) \tag{7.22}$$

$$D(T)_0 = \frac{1}{E(T)_{cs}} = \frac{1}{E(T)_g} + \sum_{i=1}^{n} \frac{1}{E(T)_i} \tag{7.23}$$

The long-term triaxial creep test results for the Quxue and Laojiaoxi dam core specimens were assumed to follow the generalized Kelvin model in the linear form of creep compliance. The trial and error method was used to find the best fit for the test results with reasonable accuracies of a simplified equation, and two Kelvin elements were taken for Eqs. (7.22) and (7.23), to give:

$$D(t, T) = \frac{1}{E(T)_g} + \frac{1}{E(T)_1}\left(1 - e^{-\frac{t}{\tau(T)_1}}\right) + \frac{1}{E(T)_2}\left(1 - e^{-\frac{t}{\tau(T)_2}}\right) \tag{7.24}$$

$$D(T)_0 = \frac{1}{E(T)_{cs}} = \frac{1}{E(T)_g} + \frac{1}{E(T)_1} + \frac{1}{E(T)_2} \tag{7.25}$$

Based on the test results, the relationships (functions) between the viscoelastic parameters and temperature are:

$$E(T)_g = 91.7e^{-0.051T} \tag{7.26}$$

$$E(T)_1 = -0.88T + 45.2 \tag{7.27}$$

$$E(T)_2 = -0.59T + 29.1 \tag{7.28}$$

$$\tau(T)_1 = 4.0 \qquad \tau(T)_2 = 8.0 \tag{7.29}$$

Long-term axial creep strain curves derived by using the SSTTC model (with the functions) are compared with test results in Figure 7.23. The curves derived using the SSTTC model fit the test results with deviations that seem acceptable at this stage of the development.

The proposed SSTTC model may be used in numerical analyses of stresses and strains in the asphalt core from the time the core is hot until it is cooled down during dam construction and loaded by lateral water pressure. The relationships between the viscoelastic properties and temperature must be defined for the asphalt mix used in the project, and the construction process and temperature changes in the asphalt core during construction and reservoir impounding must be defined. As a part of the design analysis or dam performance control, the calculated strains in the asphalt core must be compared with the tolerable strains that the core can undergo without critical cracking. If core cracking that could lead to significant leakage is likely to occur, the design of the ACED and/or the asphalt concrete mix must be modified, and mitigation measures introduced, to prevent any unacceptable consequences of the cracking.

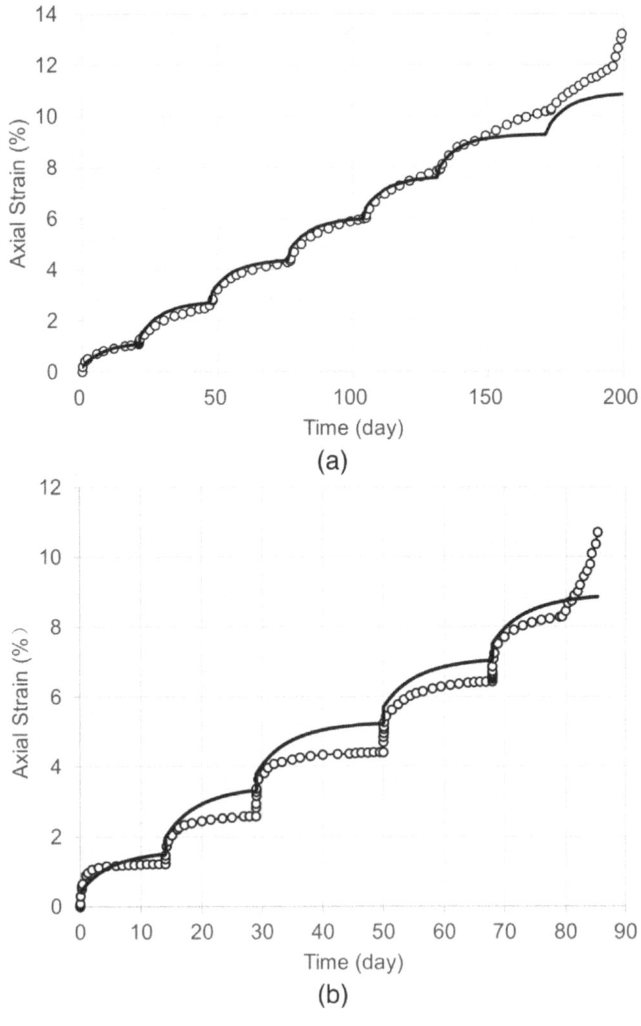

(a) Confining stress 1.0 MPa at 10°C. (b) Confining stress 0.5 MPa at 20°C.

Figure 7.23 Typical comparisons of the long-term axial creep strain curves (solid line) using the stress-strain-time-temperature creep (SSTTC) model and the test results for two different radial confining stress levels and two different temperatures.

Chapter 8

Design of asphalt core embankment dams

The design must ensure dam stability and acceptable deformations during construction, impounding, reservoir lowering and raising, flooding, and earthquake loading. The asphalt core must maintain its low permeability, high flexibility, and ductility to accommodate the deformations and strains imposed on the core by the various loading situations and potential dam foundation movements. For the Safety Evaluation Earthquake (SEE), significant cracking and damage of the core is acceptable/tolerable, but the dam design must prevent rupture and uncontrolled release of the water from the reservoir. Figure 8.1 shows an example of the design and zone designations for a typical asphalt core rockfill dam founded, for this example, on a level and sound rock foundation.

Internal erosion is not an issue for an asphalt concrete core. The main function of Zone 2 is to provide immediate, firm, and even lateral support to the slender core during both construction and dam operation. Furthermore, it is essential that Zone 2 also provides firm support to the machine paver while it is placing asphalt concrete in the core and at the same time places gravel material in Zone 2. These are essential features of the asphalt core embankment dam (ACED) construction methodology and technology. Typical stress-strain-strength parameters

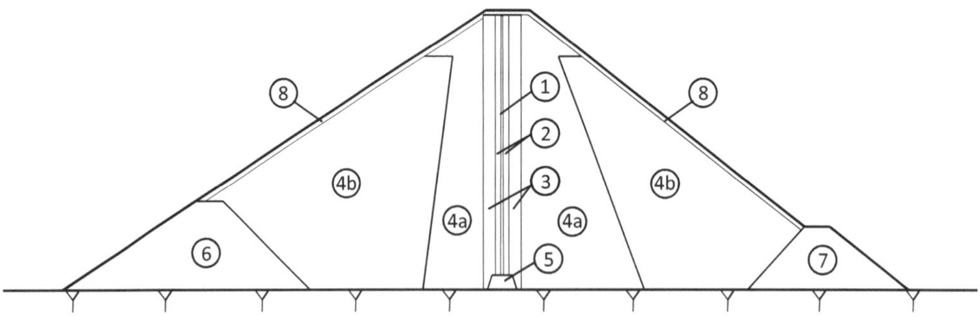

① Asphalt core; ② fine transition zone; ③ coarse transition zone; (4a) well-compacted quarried rockfill; (4b) shoulder of quarried rockfill; ⑤ concrete plinth; ⑥ upstream cofferdam; ⑦ toe drain with stable rocks and boulders; and ⑧ riprap protection.

Figure 8.1 Typical design and zone designations for an ACED on a sound and level rock foundation.

DOI: 10.1201/9781315177786-8

for the hydraulic asphalt concrete (HAC) in the core are based on test results like those presented in the previous chapters. Additional parameters required for design may be determined by applying the time-temperature superposition (TTS) principle presented in Chapter 7, which also discusses core deformations caused by long-term creep.

8.1 Asphalt core

The asphalt core wall is a flexible water barrier that must adjust to the deformations inside the embankment and to any differential settlements and displacements in the dam foundation and abutments without suffering detrimental cracking. Should cracks occur in a shear or tension zone or due to core bending deformations, the asphalt mix design should be such that viscous creep, plastic flow, and stress relaxation properties lead to crack closure and self-healing (see Chapter 5).

The asphalt core configuration and width (thickness) and the asphalt mix design are dictated by the imposed stresses and strains for the various dam loading conditions. The magnitudes of deformations and distortions may be predicted by interpreting and evaluating field performance measurements from existing dams and by carrying out numerical (e.g., finite element) stress-deformation analyses.

Asphalt core mix design is discussed in Chapters 2–4. During the suitability testing of trial mixes, laboratory specimens should be subjected to stress conditions simulating those that are anticipated in the various parts of the core in the field. The behaviour of the asphalt concrete mix is studied with respect to deformability, strength and ductility, cracking resistance and self-healing ability, shear dilatancy, and potential increase in permeability. The properties of asphalt concrete can be tailored to satisfy specific design requirements by using suitable mix proportions and grade of bitumen and by using various additives and/or reinforcing fibres in the asphalt mix as discussed in Section 3.10.

The asphalt core's position in the dam cross-section, the core geometry (vertical, inclined, or a combination), and wall width/thickness depend mainly on dam height, properties of compacted embankment materials, dam foundation conditions, and seismicity at the dam site.

ICOLD (2018) recommends that the core width (in m) at any elevation should be at least as required by using the empirical expression: Core width (in m) = 0.007 × hydraulic head (in m) on the upstream side of the core. The minimum required core width is 0.5 m. The authors recommend using a slightly larger core width as presented in the guidelines of Norwegian Water Resources and Energy Directorate (NVE):

$$\text{Core width (in m) at any elevation} = 0.50 + 0.007(H\text{-}50) \qquad (8.1)$$

where H = vertical distance (in m) from top of core to the elevation considered. All dimensions are in metres. Equation (8.1) is valid for $H > 50$ m. The minimum required core width is 0.50 m. The core width should be increased to more than that given by Eq. (8.1) at sites with significant seismicity as discussed in Chapter 9.

The core width may be allowed as low as 0.40 m if the dam is less than 15 m high at a site with low seismicity. It should be noted that the thinner the core is, the more difficult it is to ensure the minimum width of the core and to ensure the required half of the core width on both sides of the centreline. Usually, the core width is gradually reduced (tapered) from bottom to top of core, or the width may be changed in small steps of maximum 0.20 m (i.e. 0.10 m on both sides of the core centreline). Stepwise reduction in core width larger than 0.20 m may cause promote and cracks in the core during reservoir impounding and earthquake shaking and should be avoided.

However, the hydraulic gradients under the base of the foundation plinth must be limited by widening the base or by special design measures. By limiting the hydraulic gradient under the core, erosion of infill materials in rock foundation and abutment cracks may be prevented. The extent of remedial measures depends on the rock (rock mass) quality. The required base width of the plinth is specified as a function of the rock quality classification/characterization as discussed in Section 8.4. By excavating a deep core trench, one may be able to place the base of the plinth on a less weathered and jointed rock foundation.

In general, the core is placed in the central region of the dam cross-section, preferably a little upstream of the centreline to provide some extra support to the core on the downstream side. When a parapet wall is designed on the dam crest, the wall may be positioned such that the core is connected to the base of the parapet wall.

In most ACEDs built, the asphalt core is vertical from the bottom to the top of the core. However, inclined cores have also been used (e.g., Finstertal ACED, Austria, and Storvatn ACED, Norway), and finite element analyses may be carried out to study the theoretically optimum core configuration and location in the embankment considering the various static and earthquake loading conditions. In some ACEDs, the core has been made vertical except for the top 1/5 – 1/4 of the core height, which is made sloping about 20° from the vertical in the downstream direction. The "kink" is made to promote an increase in the contact normal stresses between the core and the upstream fine transition zone and to prevent any slippage or gap formation that might occur at that interface. In other cases, only the very top few metres of the vertical core have been inclined in the downstream direction, mainly to facilitate the field construction of the top narrow part of the dam (e.g., Romaine-2 Main Dam, see Section 10.5).

A vertical core from the base to the top rather than an inclined core is preferable with respect to earthquake loading that usually causes dam settlements that lead to shear and bending in a sloping core.

8.2 Transition zones

Transition zones on the upstream and downstream sides of the asphalt core are required to ensure stable and uniform lateral support to the vertical core during construction when the asphalt is hot. Transition Zone 2 on the upstream side should be sufficiently wide to accommodate differences in settlements between the core and Zone 3 during impounding. The differences may be significant if impoundment takes place after the end of dam construction (EoC) rather than during embankment construction.

Both the upstream and downstream transition zones consist of two zones, an inner finer Zone 2 with a maximum particle size of about 60 mm and an outer coarser Zone 3 with a maximum particle size of 150–200 mm (Figure 8.1). Zone 2 should be at least 1.5 m wide, and Zone 3 is commonly 3–4 m wide.

Zones 2 on both sides are placed simultaneously with the raising of the asphalt core by the machine paver to give the hot and soft core immediate lateral support. Zone 2 must be able to firmly support the paver during the placing operation and must consist of well-graded and stable material that should not be squeezed laterally into the core by the paver or later during vibratory roller compaction. Therefore, Zones 2 should preferably consist of well-graded, crushed aggregates, or a mixture of crushed and natural gravels.

The grading of the material in Zones 2 should comply with:

$$d_{50} \geq 10 \text{ mm and } d_{15} \leq 10 \text{ mm but should contain less than 5\% fines.} \quad (8.2)$$

Usually, the transition material in Zones 2 is the same on the upstream and downstream sides, but they may be specified to be different if that can improve the predicted dam performance and resistance. An example is described in Chapter 9 where the top part of Zones 2 on the downstream side is designed as a filter to arrest any migration of particles from the upstream side through cracks that may have developed during strong earthquake shaking.

If natural gravel (with sub-rounded particles) rather than crushed rock of gravel size is used in Zones 2, high lateral stresses will be imposed on the soft asphalt core during roller compaction of Zones 2. That can create excessive reduction in the core width and must be avoided. Furthermore, high lateral compaction stresses will cause the hot asphalt core to form a convex top surface with potential longitudinal cracks in the top of the core. Experience shows that such cracks, when they occur, are shallow, and they close once the next asphalt layer is placed and compacted. They are longitudinal (not transversal) and of no safety concern, but for such situations and conditions, the required minimum core width must be controlled and ensured.

The gravel material in Zone 2 is usually wetted before being placed as that facilitates compaction of the material. Dry density and porosity after compaction are commonly taken as quality control parameters. The compacted relative density is recommended to be between 70% and 75% but not much denser, as discussed and explained below.

Roller compaction of Zone 2 should occur on both sides of the core simultaneously to prevent any lateral distortion of the core while it is very hot (see Figures 1.6 and 6.14). Zone 2 should not be over compacted because that will increase the "arching" and stress transfer across the core between the upstream and downstream transition zones. If transition Zone 2 gets a compression modulus much higher than that of the compacted core, the core will tend to "hang-up" on the transition zones, and the vertical compressive stress in the core will be reduced. This reduction may cause the compressive stresses in the lower part of the core and at the interface between the core base and concrete plinth to become lower than the water pressure at full reservoir level. Then, hydraulic splitting (separation) may occur at the contact. To study this phenomenon, pressure cells mounted on top of the plinth under the asphalt core have been used to measure the vertical contact stresses between the base of the core and the plinth. Measurements made by the authors showed in one case that the vertical stresses under the asphalt core were as low as 50% of the vertical stress corresponding to the weight of the asphalt core and fill material above the pressure cells. In another case, the vertical stress was measured to be about 85% of the vertical stress. In both cases, the measured values of contact stress under the base of the core remained nearly constant with time after end of construction.

Charles (1998) discussed this issue and presented a summary of the relatively few cases of ACEDs where such contact stress measurements had been made. The review shows that the contact stresses under the base of the asphalt core on the plinth varied from about 50% to 90% of the stress corresponding to the weight of core and fill material above the base. A contact stress higher than the hydrostatic pressure at that level prevents hydraulic separation from occurring.

The stress values measured by pressure cells imbedded in an earth core or in granular filters in the field or in the laboratory often seem to be unrealistic and unreliable and have been disregarded in most studies of stress measurements in embankment dams. Much of the uncertainty seems to stem from the soil disturbance and seating problems for the pressure cells during installation inside the embankment fill. In addition, there is a difference in compressibility between the pressure cell and the soil it replaces. However, pressure cells mounted in the top of the concrete plinth under the core in an ACED should give realistic values for the contact stress and have been relied upon in these studies of arching across the asphalt core.

If hydraulic separation should occur in the field at the core-plinth interface, it will lead to a significant increase in leakage and should easily be detected by seepage monitoring. No ACED has been identified where there has been hydraulic separation in or under the core.

Non-linear finite element stress-deformation analyses have been applied to study the arching and redistribution of vertical stresses among the various zones in the core region. A challenge in these analyses is to evaluate the input value for the asphalt core compression modulus which, as discussed in Chapter 7, is stress, time, and temperature dependent. The ratios between the compression moduli for the core, Zone 2, Zone 3, and inner part of Zone 4 govern the stress distribution among the various zones in the core region (Figure 8.1).

To reduce the friction "hang-up" between the core and transition Zone 2, and thereby, increase the vertical stresses at the core base, Wang et al. (2018) installed a thin "sliding layer" between the core and Zone 2. Laboratory tests showed that the interface friction stresses were reduced by about 40% when a 20-cm thick "sliding layer" of sub-rounded natural sand rather than angular crushed sand was used. More investigations of the pros and cons and the practicality of using a design with such a "sliding layer" are in progress.

The particle sizes for aggregates in the asphalt core, Zone 2, Zone 3, and shoulders should be compatible. The following guideline is presented by ICOLD (2018):

$$d_{100} \text{ core} \geq d_{10} \text{ trans, and } d_{100} \text{ trans} \geq \tfrac{1}{4}\, d_{100} \text{ shell} \tag{8.3}$$

Note that for ACEDs used for daily pumped-storage reservoirs where the raising and drawdown of the reservoirs are frequent and are occurring at a high rate, filter criteria may be required between the upstream shoulder and the transition zone to maintain the stability of the transition zone during rapid drawdown.

8.3 Dam shoulders

When the reservoir is impounded, the water pressure causes the core wall to displace and deflect, but it is restrained by the transition zones and shoulders, and bending moments develop in the core. As shown in Figure 8.1, there is an especially well-compacted shoulder zone on the downstream side of transition Zone 3 to prevent the core from undergoing excessive lateral displacements during impounding and earthquake loading. During the early phase of impounding after the dam is completed, the top of a vertical core may tend to deflect a little in the upstream direction due to the submergence and wetting settlements of the upstream shoulder. However, for higher reservoir levels, the core bends in the downstream direction as shown in the examples in Sections 10.2 and 10.7.

For static loading conditions, the stability analyses for the dam slopes are traditionally performed by the classical limit equilibrium method (LEM) to calculate the factor of safety (FoS) against slope instability. The stability is mainly governed by the shear strength properties of the compacted shoulder material and, for deeper failure surfaces, some of the material in the transition zones. The potential sliding surface may, in some cases, cut through a weak foundation layer under the shoulder. That may change the critical failure mode and cause a significant reduction in the computed FoS for the dam slope.

The trend is to supplement the LEM analyses with finite element method (FEM) analyses to estimate the stress distribution and deformations inside the dam body, in the foundation, and in the core region, as the dam is constructed in compacted layers. The FoS may also be computed by using the stress-deformation FEM analysis by employing the "strength reduction method". This is achieved by studying the increase in displacements, and any changes in displacement pattern as the shear strength parameters are gradually reduced in the FEM analysis (Griffith and Lane, 1999). The FoS values computed by the two methods (LEM and FEM) should be equal or very similar and provide a check on the reliability of the computations. Recent practice replaces the LEM method with the use of the FEM method that gives more information about the slope stability and deformation pattern.

Traditionally, only 2-dimensional stability analyses have been performed, but 3-dimensional analyses for FoS and for deformation computations are possible with modern computer software and are used when there are uncertainties about the effects of the geometry and boundary conditions on the computed results (see Section 10.8).

In general, the embankment slopes for an ACED can be made somewhat steeper than for an earth core embankment dam (ECED) with a central core as most of the impervious earth material in a wide earth core is replaced by material with higher shear strength and stiffness (rockfill, gravel, and alluvium) in the ACED.

For a well-compacted embankment of good rockfill on a sound bedrock foundation, the dam slopes may be as steep as 1V:1.3H as demonstrated by the Finstertal ACED in Austria and Storvatn ACED in Norway and many other dams with good and well-compacted rockfill. Even inside these dams with very steep slopes, the vertical and horizontal displacements are measured to be small with shear strains in the asphalt core well below tolerable values and do not cause detrimental fissuring or cracking (see Chapter 3).

Case studies have also demonstrated that ACEDs may successfully be built with lower quality rockfill than used in the early ACEDs. The Feistritzbach ACED in Austria (Tschernutter, 1997) and the Storglomvatn and Holmvatn ACEDs in Norway are examples where weak, weathered rockfill materials were used. ACEDs have also been built where rockfill was not available, but alluvium of various gradations and quality was used. If the embankment is built of sand and gravel mixtures or with weathered materials or rock debris, the friction angle can be significantly lower than for fresh, quarried rockfill, and gentler dam slopes must be used.

For high ACEDs built with shoulder material of soft rock and low deformation modulus, the slope stability may well be satisfactory, but the magnitude of deformations in the dam and lateral displacements of the core must be considered and analysed. A well-compacted gravel zone downstream of Zone 2 may be required to restrain the downstream displacements and bending of the asphalt core during reservoir impoundment and potential earthquake loading. An example of such a design with compressible shoulder materials is the 89-m high Jinfeng ACED in Sichuan Province, China, which was constructed using soft sandstone debris (Liao et al., 2023). The wet, unconfined compressive strength (UCS) of the soft sandstone was in the range 12–15 MPa, and the friction angle of the sandstone was low. Gentle upstream and downstream dam slopes of about 1.0V:2.2H were used as the dam is located in a region with some seismicity. To limit the lateral displacements and bending of the core, a 32-m high, well-compacted gravel zone (wall) behind Zone 2 restrains the downstream lateral displacements and bending of the asphalt core. The Jinfeng ACED has performed well since the completion of the dam in 2020.

The foundation conditions under the dam shoulders must be carefully evaluated when analysing the dam slope stability if the embankment is founded on layered soil or strongly weathered and jointed rock. The foundation may, for instance, contain weak layers or bedding planes that may develop excess pore pressures and reduced shear strength during and after earthquake shaking.

8.4 Optimal dam core alignment

Once the general location of the dam axis has been selected, attention focuses on the foundation preparation required for the various potential axis locations and orientations. If possible, a short dam axis across the valley is used to reduce the volume of the dam embankment. However, the location and design of the spillway structure are also essential considerations and to avoid riverbeds and dam abutments with very irregular topographical features under the plinth. Very steep abutments, more than 60°, may require special core-abutment interface design to ensure sufficiently high normal contact stresses between core and plinth to avoid interface leakage and potential slippage (Section 8.8).

The rock foundation for the plinth should be sound and erosion resistant, should not cross any fault zones, and can be grouted. If weak, jointed compressible rock is encountered locally along the plinth alignment, the concrete plinth may be designed to bridge over the weak zones. However, foundation erosion and leakage under the "bridge" must be prevented. The dam axis should be located where the compressible overburden (e.g., alluvium) or depth of strongly weathered rock is shallow; otherwise, a deep cut-off trench or a cut-off wall may be required. The concrete plinth can be placed on top of the cut-off wall as has been done on several of the existing ACEDs and is discussed in Section 8.5.

The dam axis may be given a slight upstream (convex) curvature to promote a compressive, cross-valley stress in the core, and sometimes done mainly for aesthetics. For the 128-m high Storglomvatn ACED with a dam crest length of 825 m, the dam axis has a slight upstream curvature with a radius of 1,500 m (Figure 1.1). The dam axis of the 45-m high Riskallvatn ACED and the 52-m high Styggevatn ACED, both dams in Norway, and the 60-m high Greater Ceres ACED in South Africa have all been built with slight upstream curvatures.

At the site of the 100-m high Storvatn ACED, Norway, with a dam crest length of 1,472 m, the dam axis giving the minimum dam volume is partly straight, partly curved concave, and partly convex to the reservoir. A straight axis across the valley would have given a maximum height of the dam approximately 10 m higher than the final dam axis selected and a corresponding dam volume of about $10^6 m^3$ larger. For the dam axis to be everywhere curved convex to the reservoir, the difference in volume would be larger. Avoiding the local concave curvature of the dam would have required an extra cost of approximately 10%. The question was therefore how large negative effects (i.e., increased probability of unsatisfactory performance) a concave curvature would have on the dam behaviour. It is known from field observations and theoretical analyses that the top part of a central core in a rockfill dam usually displaces a little in the upstream direction during the early stages of reservoir impoundment due to settlements in the submerged upstream shell. However, the core is subsequently pushed in the downstream direction as impounding proceeds. An upstream displacement in a core with convex curvature creates extensional strains and may possibly lead to transverse cracks in the dam core. Finite element analyses showed, however, that the extensional strains to be expected for the Storvatn ACED would be very small, and the convex curvature was found to be acceptable. Over a portion of the dam length towards the left abutment, the core was given a concave curvature with a radius of 300 m to save significant construction costs. Therefore, an alignment of the dam with sections of both convex and concave curvature was accepted. The dam has shown very satisfactory performance since its completion in 1987.

Another example is Finstertal ACED in Austria, with a dam crest length of 652 m. It straddles a ridge to reduce total dam volume and costs (see Section 10.2). An optimum dam axis location with respect to embankment volume and stability was achieved with a slight upstream (convex) curvature of the main body and a short concave curve towards the right abutment (Pircher and Schwab, 1988). The Finstertal ACED has shown very satisfactory performance since it was impounded in 1980.

8.5 Plinth design and construction

The foundation area to be prepared and treated under the core and the fine transition zones for an ACED is much smaller than for an earth core option with filters (ECED) at the same site. The concrete plinth under the asphalt core may be seated directly on the riverbed rock foundation and abutments, in a deep trench, or on top of a cut-off wall going through the pervious foundation down to or partway to an impervious layer or bedrock (see Section 8.6). An important purpose of the plinth is to serve as a cap for the contact and curtain grouting under the core, and the concrete plinth should be anchored into the rock by bolts to allow high grouting pressures. The plinth is lightly reinforced but has sufficient reinforcement to prevent shrinkage cracks, and the anchor dowels may be connected to the longitudinal plinth reinforcement to ensure sufficient pull-out resistance for the anchors.

The thickness of the plinth depends on the foundation profile and the degree of irregularity of the rock foundation and abutments. The plinth should have a minimum concrete thickness of about 0.4 m. In some locations along the plinth, the thickness of concrete in and under the plinth may turn out to be quite large to achieve a relatively smooth and regular design profile of the top surface of the plinth as the base for the asphalt core. To obtain the optimum foundation profile, a combination of excavation (e.g., blasting) and infilling with concrete is required depending on the local geological and topographical conditions. An optimization analysis is required to find the best solution. The quality of concrete used for infilling under the plinth may be lower than in the plinth itself. Not only should the core rest on the plinth but also the fine transition Zone 2 on both sides of the core. As an example, Figure 8.2 shows the upper part of the core trench with plinth for the 167-m high Moglicë ACED in Albania.

The base width of a concrete plinth resting on a rock foundation is governed by the dam height, the rock conditions under the plinth, the plinth structural stability, and the method of rock foundation treatment. The allowable hydraulic gradient under the plinth depends on the rock quality (degree of jointing, fracturing, and weathering). The width usually decreases along the abutments from the bottom to the top of the dam according to the hydraulic head unless the foundation characteristics are much inferior towards the top of the abutment. Guan (2014) suggested that allowable hydraulic gradients under the plinth may be larger than or equal to 20 (for fresh or slightly weathered rock), 10–20 (moderately weathered), 5–10 (intensely weathered), and 3–5 (fully weathered). The allowable hydraulic gradient is given as 20 (sometimes more) for a fresh or slightly weathered rock foundation with few joints or fractures. A rock mass quality classification/characterization system, e.g., the GSI system, may be used (Hoek et al., 1995). Cruz et al. (2009) showed, as an example, the design recommendations for the very high Barra Grande CFRD in Brazil, with allowable hydraulic gradients under the plinth between 12 and 19.

Figure 8.2 Moglicé Dam, Albania under construction. (Courtesy of WALO International AG-Department-Hydraulic Asphalt Engineering.)

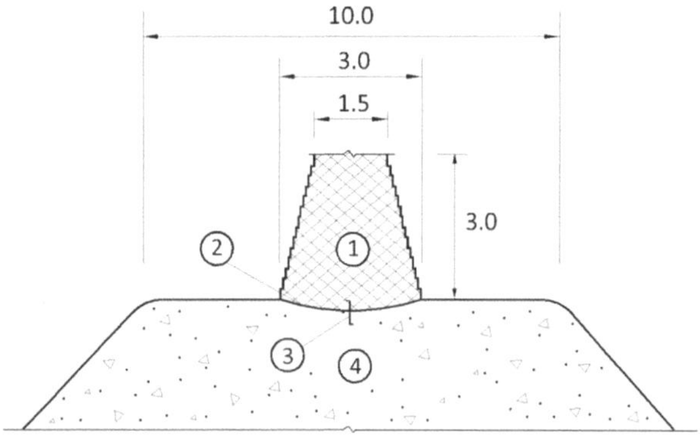

① Asphalt core; ② 1.0–2.0 cm thick asphalt mastic; ③ copper water stop; and ④ concrete plinth on top of bedrock (all dimensions in metres).

Figure 8.3 Details of core-plinth connection on rock above the ground level for the 174-m high Quxue ACED, China (see Figure 10.23 and Zone 5 in Figure 10.24).

Figure 8.3 shows the asphalt core – concrete plinth connection for the 174-m high Quxue ACED in China. The figure shows the situation where the rock surface is above the ground level, and no foundation trench was required. The asphalt core is widened (flared) against the foundation plinth to twice the core width. The length of the widened (flared) section should be at least–3 times the core width.

Traditional practice has been that at the bottom of the core on the foundation plinth, one placed two layers of asphalt concrete (20 cm compacted layer thickness) with widths twice that of the core width. Above that, two layers were placed of widths 1.5 times the core width to build a base for the core. In more recent ACEDs, a more elaborate construction procedure has been followed to achieve a more gradual widening of the core against the foundation plinth (as shown in Figure 8.3). The height of the widened (flared) section should be 2.5 times the core width. The core width for the Quxue ACED is 0.60 m at the top and 1.50 m at the bottom.

The concrete plinth is commonly built in 6–10 m long segments depending on the foundation conditions and rock profile under the plinth. The segments are separated by construction joints sealed with water stops made of rubber, copper, or high-density foam embedded in the joints. The plinth along the foundation and abutments should be designed in such a way that angle changes between adjacent (abutting) plinth segments should be not more than 35°. The purpose of that recommendation is to limit the shear distortions in the asphalt core wall caused by a geometrically irregular foundation profile.

For some ACEDs, the plinth has been built without construction joints, and the steel reinforcement has been continuous. As stated above, the plinth is generally lightly reinforced, only sufficient to prevent shrinkage cracks. It is a good practice to apply a layer of asphalt mastic on the upstream side of the plinth to seal any cracks that may develop. Application of heavy industrial sealing tape over the joints on the plinth surface has also been used.

Some designers make a shallow concave profile in the top of the plinth to make it easier for the core to slightly pivot at the interface between core and plinth and at the same time keep the interface watertight (example in Figure 8.3).

8.6 Plinth on cut-off wall through pervious foundation

The plinth may be built on top of a cut-off wall going through or partway through a pervious foundation under the plinth. The structural connections between the cut-off wall under the plinth, the concrete plinth, and the asphalt core above the plinth must be stable and watertight. There will be a difference in vertical settlement between the plinth on top of the cut-off wall that goes down to firm ground or bedrock and the more compressible soil overburden or weathered rock on either sides of the wall. A differential settlement results in downward friction forces (down drag) on the asphalt core and cut-off wall and high stresses in the connection between the dam core and the plinth.

The cut-off wall will also undergo lateral displacements and differential lateral displacements in the downstream direction when the reservoir is impounded. The design must keep the cut-off wall watertight without detrimental cracking. Movements of the cut-off wall causes movements of the plinth and potential shear distortions in the asphalt concrete core.

Many designers have attempted to adjust the compression modulus of the material in the cut-off wall such that the vertical compression of the cut-off wall will be the same as the compression (settlement) of the soil layer or weathered rock adjacent to the cut-off wall. That is not easy to achieve because the vertical strains may vary significantly along the dam across the valley, under the dam, and over the height of the cut-off wall. Finite element analyses may assist in the evaluation of the differential vertical settlements and the transfer of stresses from the surrounding ground to the cut-off wall. Likewise, finite element analyses may be used to estimate the lateral displacements of the cut-off wall in the downstream direction and the potential cracking that may occur where the cut-off wall approaches the abutments. The material used in the cut-off wall has in many cases been "plastic concrete", which is basically regular concrete with addition of bentonite in the mix to obtain a compression modulus for the wall that will give vertical compression strains in the wall approximately equal to the vertical strains in the overburden on either sides of the wall. This reduces the transfer of stress from the surrounding ground to the cut-off wall.

Another way of reducing high compression stresses in the cut-off wall is to install a compressible element in the connection between the asphalt core and the plinth as shown in the examples presented below. The following figures in this chapter show examples of plinths built on top of deep cut-off walls going through pervious and compressible dam foundations.

Many earth core embankment dams (ECEDs) have the core built on top of impervious cut-off walls (e.g., ICOLD, 2005). The mitigating measures to prevent large drag-down forces are basically the same as those discussed for asphalt core dams. In addition, for the ECED, the effects of various lengths of cut-off wall penetration into the earth core (overlapping of the core and cut-off wall) have been studied to obtain the optimal connection for the specific site and loading conditions.

Figures 8.4 and 8.5 show the structural connection between the asphalt core and the concrete cut-off wall for the 125-m high Yele ACED in China. It may be noted that for the Yele ACED, an upstream impervious geomembrane was installed on the foundation covering the area from the cut-off wall to the upstream dam toe. Detailed descriptions and field measurements are presented in the case study in Section 10.4.

Figures 8.6 and 8.7 show the structural connection between the asphalt core and the concrete plinth for the 78-m high Xiabandi ACED in China, with a 148-m deep overburden (Qin et al., 2011).

Figure 8.8 shows the structural connection between the asphalt core and concrete plinth for the 75-m high Daheyan ACED in China with a 186-m deep overburden (Li et al., 2021). The cut-off wall goes through the pervious soil overburden and into the rock foundation and is the deepest cut-off wall built in China to date.

Figure 8.9 shows the connection between the asphalt core and the compressible clay-cement cut-off wall for the 28-m high Eberlaste ACED in Austria constructed during the period 1966–1968 on a very compressible and non-uniform alluvial foundation.

For the Daheyan ACED in Figure 8.8, the designer decided to not install any compressible layer/element on top of the cut-off wall, and the design appears to work well.

8.7 Foundation gallery

Many of the early ACEDs were built with a gallery in the rock directly under the core or beside the core base, as for the High Island East Dam (105 m, Hongkong, 1978); Grosse Dhünn Dam (63 m, Germany, 1980); Finstertal Dam (100 m, Austria, 1980); Wupper Dam (40 m, Germany, 1986); and Feistritzbach Dam (86 m, Austria, 1990). The purpose of installing a gallery close to the base of the core is to facilitate grouting of the foundation during and after embankment construction, for use in inspection and drainage control, and for instrumentation and monitoring.

Constructing a gallery may result in a significant cost increase. In many cases, the main advantage of having a gallery is that it gives the possibility of post-construction grouting if that is deemed necessary. Another potential advantage is to shorten the dam construction period by separating the embankment construction activities and foundation grouting activities, which can occur independently when a grouting gallery is constructed. However, if the dam seats on a relatively sound rock foundation, constructing a gallery could damage the rock by additional fracturing due to blasting and could prolong, rather than shorten, the construction period. None of the

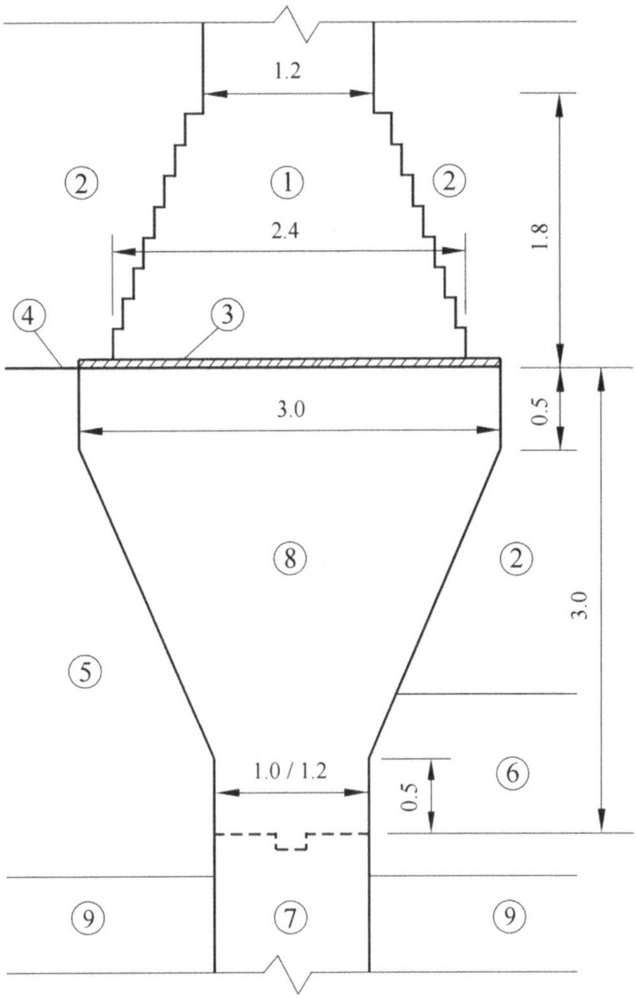

The left side of the figure is upstream. ① Asphalt core; ② transition zone; ③ 1–2 cm thick sandy asphalt mastic (SAM); ④ geomembrane covering foundation to upstream dam toe; ⑤ silt; ⑥ filter and drainage layer; ⑦ concrete cut-off wall; ⑧ reinforced concrete plinth; and ⑨ foundation overburden.

Figure 8.4 Structural connection between the asphalt core and concrete plinth for the 125-m high Yele ACED, China (all dimensions in m).

170 Norwegian large embankment dams on fairly good rock foundations have such a gallery. For the 128-m high Storglomvatn ACED (case study in Section 10.3), there is no gallery under the asphalt core in the left abutment and riverbed. However, a tunnel was built at 25–40 m depth in the right abutment to facilitate grouting during construction as the right abutment is in a karstic rock region. The gallery built under the Storglomvatn ACED would have made it possible to undertake post-construction grouting. However, up to now, that has not been found to be necessary (see Section 10.3).

Figure 8.5 Concrete plinth on top of the cut-off wall for the 125-m high Yele Dam, China.

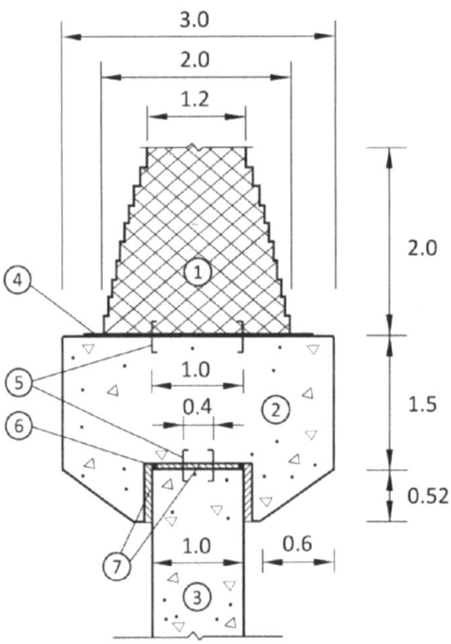

① Asphalt core; ② concrete plinth; ③ concrete cut-off wall; ④ 1.0–2.0 cm thick SAM; ⑤ copper water stop; ⑥ expansion seal (cross-section 20 mm × 30 mm); and ⑦ 20 mm thick asphalt flooring tile (all dimensions in m).

Figure 8.6 Structural connection between the asphalt core and concrete plinth for the 78-m high Xiabandi ACED with 148-m deep overburden. (Courtesy of Hui Pang, Shaanxi Province Institute of Water Resources and Electric Power Investigation and Design.)

Figure 8.7 Concrete plinth for the 78-m high Xiabandi ACED, China, on top of a 148-m deep cut-off wall. (Courtesy of Hui Pang, Shaanxi Province Institute of Water Resources and Electric Power Investigation and Design.)

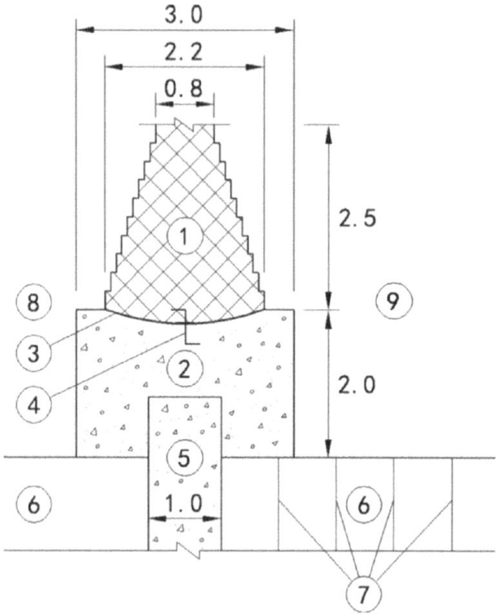

① Asphalt core; ② concrete plinth; ③ 1.0–2.0 cm thick SAM; ④ copper water stop; ⑤ concrete cut-off wall; ⑥ foundation overburden; ⑦ four rows of blanket grouting with depth 10–12 m; ⑧ upstream side; and ⑨ downstream side (all dimensions in m.)

Figure 8.8 Structural connection between the asphalt core and the cut-off wall for the 75-m high Daheyan ACED, China with a 186-m deep cut-off wall. (Courtesy of Huaxin Huang, Hunan Water Resources and Hydropower Survey, Design, Planning and Research Co. Ltd.)

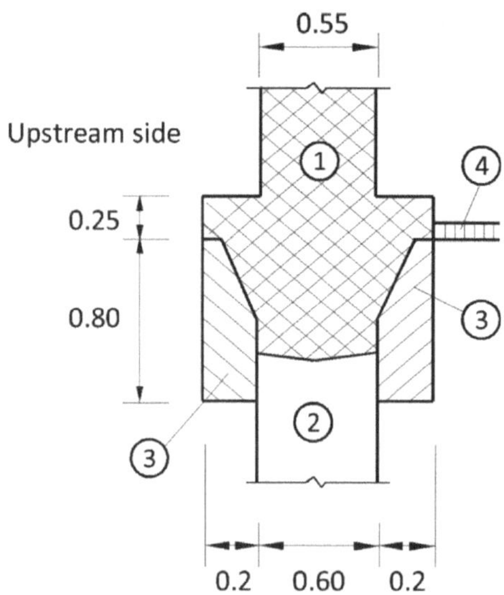

① Asphalt core; ② clay-cement slurry cut-off wall; ③ asphalt mastic; and ④ asphalt gravel layer (all dimensions in m).

Figure 8.9 Connection between the asphalt core and the compressible clay-cement cut-off wall for the 28-m high Eberlaste ACED, Austria, built in 1966–1968 as the first ACED with plinth on top of cut-off wall (ICOLD, 2018).

In recent years, many high ACEDs have been built without constructing accompanying galleries, e.g., Romaine-2 Main Dam (112m, Canada, 2013); Zarema Dam (153m, Ethiopia, 2017); Dashimen Dam (129m, China, 2019); Bamudun Dam (129m, China, 2020); and Niya Dam (134m, China, 2023).

However, for geologically complex foundation conditions, galleries or tunnels may be required to be able to perform additional foundation investigations and/or to facilitate the construction of deep, underground seepage barriers. Examples of some large ACEDs with complex and challenging foundation conditions are presented below. When there are fault joints and karst cavities in the abutments, or the banks consist of deep overburden, galleries or tunnels have been built in the abutments at various depths. The galleries are mainly used for foundation grouting and cut-off wall construction, as described in the case studies for the Yele and Quxue ACEDs (see Sections 10.4 and 10.7).

The grouting for the bottom part of a deep overburden under a concrete cut-off wall may be carried out from a gallery. The Huangjinping ACED (85.5m high with an overburden depth of 134m, China, 2015) is such a case. Figure 8.10 shows the gallery between the asphalt core and the concrete cut-off wall. The gallery is a large concrete monolith on top of the cut-off wall across the river channel to connect with the galleries used for construction of the grout curtains in the rock abutments. To limit crack formation, the gallery was built in segments, and the joints between the segments were designed with water stops. For the Huangjinping ACED, the entire surface of the gallery was covered with a 10-mm thick flexible bitumen membrane as shown in Figure 8.10. Large differential settlements were anticipated between the gallery and the right

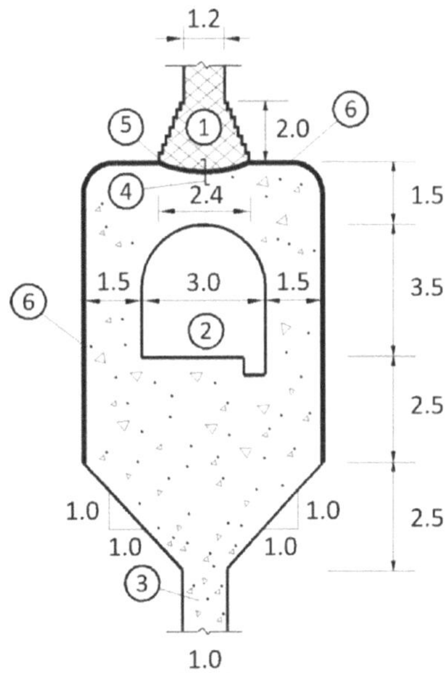

① Asphalt core; ② grouting gallery; ③ concrete cut-off wall; ④ copper water stop; ⑤ 1.0–2.0 cm thick SAM; and ⑥ 10 mm thick modified bitumen membrane covering the entire surface of the gallery (all dimensions in m).

Figure 8.10 Gallery under asphalt core on concrete cut-off wall for the 85.5-m high Huangjinping ACED, China, with a 134-m deep overburden. (Courtesy of Shunbin He, POWERCHINA Chengdu Engineering Co. Ltd.)

rock abutment, and a SAM (sandy asphalt mastic) layer was placed on the bottom of the segment of the gallery resting on the right rock abutment.

The gallery of the Suwalong ACED (110 m high with an overburden depth of 91 m, China, 2021) is like the gallery of the Huangjinping ACED, but it was decided not to build the gallery in segments. When the dam embankment was about 45 m high, cracks and leakages occurred in the gallery near the right rock abutment. The cracks were chemically grouted, and the leakages were stopped. However, after the dam was completed and the reservoir was impounded, the sealed cracks opened again, new leakages occurred, and additional grouting was required.

For some sites, constructing a gallery under the core and above the cut-off wall may further complicate the deformation situation, and the gallery may rather be put adjacent to the connection between the asphalt core and the concrete plinth. The Xiabandi ACED, China, (78-m high, overburden depth 150 m, 2009) and the Pangduo ACED, China, (80-m high, overburden depth 150 m, 2014) are two such cases where this unconventional solution was selected. Figure 8.11 shows the structure of the gallery and the connection between the asphalt core and the concrete cut-off wall for the Xiabandi ACED.

Due to the concern that constructing a gallery would further complicate the deformation situation between the asphalt core and cut-off wall, the gallery was located 11.7 m downstream from the asphalt core axis for the Yele Dam, as shown in Figure 8.12. The case study is presented in Section 10.4.

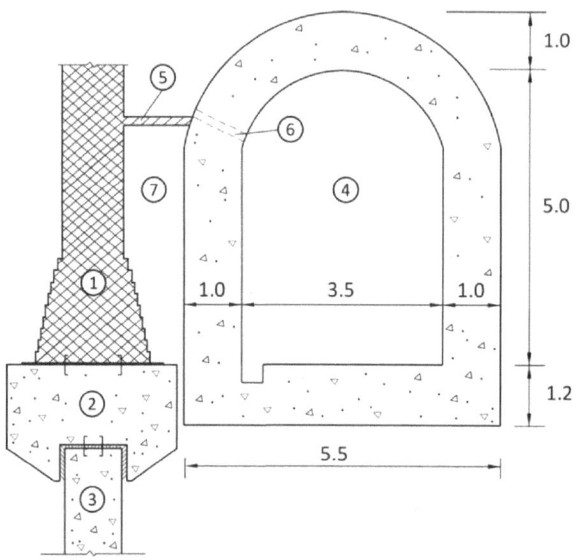

① Asphalt core; ② concrete plinth; ③ concrete cut-off wall; ④ grouting gallery; ⑤ 40 mm thick asphalt concrete; ⑥ drainage hole; and ⑦ fine gravel (all dimensions in m).

Figure 8.11 Gallery and connection between the asphalt core and concrete cut-off wall for the 78-m high Xiabandi ACED, China, with 148-m deep overburden. Details of structural connection between asphalt core and concrete plinth are shown in Figure 8.6. (Courtesy of Hui Pang, Shaanxi Province Institute of Water Resources and Electric Power Investigation and Design.)

Figure 8.12 Gallery located 11.7 m downstream from the asphalt core axis for the 125-m high Yele Dam with a 55–160 m deep overburden below the riverbed.

8.8 Design of core interface for very steep rock abutments

If the slope of the abutment is steeper than about 60°, extra measures should be taken to increase contact stresses between the asphalt core and the plinth and avoid the opening of leakage paths at the interface. The interface between the core and the plinth should be made to incline inward

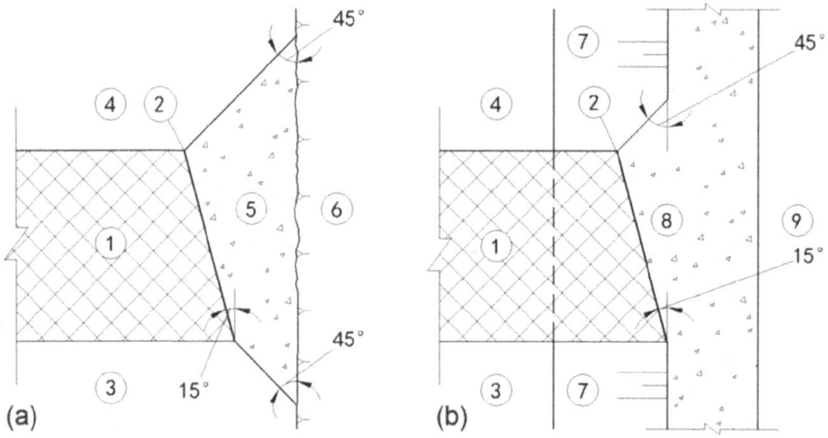

① Asphalt core; ② 1.0–2.0 cm thick SAM; ③ upstream; ④ downstream; ⑤ concrete plinth; ⑥ right abutment; ⑦ concrete wall; ⑧ block-out in the wall; and ⑨ spillway or power house.

Figure 8.13 Schematic diagrams of cross-sections of connection between the asphalt core and (a) steep abutment and (b) concrete structure.

about 15° towards the centreline of the valley to converge in the downstream direction as shown in Figure 8.13a. This inclination increases the interface contact stresses when the water pressure from the reservoir acts on the core. The inclined part should also extend under the fine transition Zone 2. For extra insurance against leakage, water stops may be installed at the core-abutment interface.

For the contact between the core and the steep wall of a spillway or powerhouse a block-out in the concrete wall may be the best way to achieve the 15° inclination and the increase in contact stresses as shown in Figure 8.13b. An example of the application of this design solution is the Foz de Chapeco ACED in Brazil. The performance of the dam has been very satisfactory since it was built in 2003. For the interface with spillway or powerhouse, drilling and interface grouting may take place from the inside of the concrete structure if that is required.

A SAM (sandy asphalt mastic) layer placed between the core and the plinth can effectively increase the ability to accommodate shear and extensional deformations of the core-plinth connection without cracking (Wang et al., 2017a). The asphalt mastic with some added styrene-butadiene-styrene (SBS) block copolymer or other additives can also improve the core-plinth connection performance. Water stops embedded in the plinth and core can further improve the connection and make it watertight. One row of copper water stops is used in the core-plinth connection for the asphalt core in the 174-m high Quxue ACED (see Section 10.7), and two rows of copper water stops are used for the 78-m high Xiabandi ACED (Figures 8.6 and 8.7). The Xiabandi Dam is founded on an overburden 148 m deep.

Chapter 9

Design features to increase earthquake resistance

9.1 Introduction

The recent book edited by Hinks (2023) covers most aspects of earthquake engineering for dams and reservoirs. Several chapters are devoted to various types of embankment dams, their behaviour during earthquakes, and the damage that may occur if the dam and appurtenant structures are exposed to earthquake loading.

Embankment dams undergo volumetric compaction, crest settlements, and slope shear displacements when subjected to earthquake shaking. The slender asphalt concrete core must be designed in such a way that it can accommodate the dam deformations caused by the cyclic and permanent strains and distortions imposed by earthquake shaking. The dynamic material properties of the core itself, e.g., shear modulus and damping properties, have little effect on the overall dynamic response of the dam. The core must be designed to follow the deformations imposed by the embankment without undergoing detrimental cracking and excessive leakage. Hydraulic asphalt concrete (HAC) in the dam core has the properties to satisfy these requirements if the asphalt mix is properly designed, thus giving the core flexibility and ductility.

9.2 Dynamic analysis of embankment dams

No high asphalt core embankment dam (ACED) has yet been exposed to significant earthquake loading. The design of new ACEDs and the safety evaluation of existing ones in seismic regions must therefore be based on theoretical analyses, model testing, and field performance observations of other types of embankment dams subjected to earthquake shaking, such as presented by Seed (1979), Longtin et al. (2013), Fell et al. (2015), Han et al. (2016), Jiao et al. (2020), Mejia (2021), and Hinks (2023).

Extensive material testing of asphalt concrete has been undertaken to investigate the core's resistance to cracking for various loading conditions, and that information is used for the design of new ACEDs and when evaluating the performance and safety of existing dams.

The dynamic stresses and strains imposed by an earthquake on a central core of earth or asphalt concrete, or an upstream facing, have been analysed and discussed by several authors, e.g., Belloni et al. (1988), Le Coroller et al. (1988), Gazetas and Dakoulas (1992), Ghanooni and Roosta (2002), Jafarzadeh and Javaheri (2003), Li et al. (2003), Wieland (2003), Yasuda et al. (2003), Baziar et al. (2003, 2006), He et al. (2008), Akhtarpour and Khodaii (2009), and Feizi-Khankandi et al. (2009).

Dynamic numerical analyses of embankment dams subjected to earthquake loading may be performed by means of commercially available software that uses various forms of non-linear

DOI: 10.1201/9781315177786-9

elastic and simplified elastic-plastic constitutive material models for the various embankment materials. The analyses are used to predict the dynamic stress and displacement response during the earthquake shaking as well as to predict the permanent (residual) settlements and horizontal displacements resulting from the shaking (USSD, 2022). The computed deformed shape of the asphalt core shows where extensional strains and shear strains may lead to cracking and possibly local rupturing of the asphalt core, and where a second line of defence may be required, for instance, by using a geo-membrane in a critical area.

The dynamic properties of the shell (shoulder) and transition materials used in the dynamic analyses, like shear modulus and damping coefficient as function of strain level, may be obtained by performing special laboratory tests or by using available published experimental results and field observations (e.g., Yasuda et al., 2003; Han et al., 2016; Menq and Stokoe, 2003; Stokoe et al., 2017; Hinks, 2023). Dynamic deformation and material damping properties of HAC have also been investigated and are presented by Feizi-Khankandi et al. (2008), Wang and Høeg (2011), Akhtarpour and Khodaii (2013), Han et al. (2023), and Meng et al. (2023).

In general, the dynamic analyses show that when the embankment dam is founded on a rigid rock foundation, the largest dynamic amplifications of the earthquake ground accelerations occur in the top part of the dam where extensional and shear strains may cause cracking and leakage. Non-linear numerical analyses that have been performed of embankment dams on rigid rock foundations show that below about 25% of the dam height from the crest, the asphalt ore is predicted to be predominantly in compression, but tensile stresses may develop in the core towards the abutments.

For embankments on compliant foundations, e.g., soil, soft rock, or strongly weathered rock, the maximum dynamic amplifications of the foundation accelerations occur at lower levels in the dam, closer to the dam-foundation interface. Large displacements may then occur towards the bottom of the core, and the dam slopes may have to be anchored to the dam interior by earth reinforcement or anchors as suggested by Seed (1979) based on dynamic analyses and field observations of dam slope movements. An example of stabilising the slope by tension anchors is the case study of Yele ACED in China described in Section 10.4.

Dynamic analyses must be carried out for each case to determine critical levels for the dam core and for the dam slopes, and to evaluate potential structural mitigation measures.

Due to the complexity of non-linear dynamic analyses, the computed residual embankment slope displacements should be compared with the displacements estimated by the Newmark type simple sliding block (wedge) approach (Newmark, 1965; Bray and Travasarou, 2007; Bray and Macado, 2019; Hinks, 2023) and by empirical relationships for crest settlements based on field observations (e.g., Swaisgood, 2014). It should be noted that the Newmark method of displacement analysis, which is much used, only considers shear (deviatoric) strains and does not include volumetric strains that cause additional crest settlements and loss of freeboard.

For severe earthquake loading, numerical analyses and model testing of ACEDs show significant settlements in the transition Zone 2 upstream and downstream of the core towards the crest, while there are only very small settlements in the asphalt core itself. The upstream side of the crest shows settlements much larger than the settlements on the downstream side. Furthermore, the computed horizontal component of the upstream displacement on the upstream side is significantly larger than the horizontal component of the upstream displacement on the downstream side, thus creating extensional strains in the crest. Similar behaviour is observed in shaking table and dynamic centrifuge tests (e.g., Baziar et al., 2009; Feizi-Khankandi et al., 2009; Jiao et al., 2020). Thus, deformations in the top part of the dam due to static and cyclic loading, may cause longitudinal cracks to develop in the dam crest. This cracking is not an immediate safety concern as the cracks are longitudinal (not transversal) and shallow. However, they should be repaired after the first full impoundment or the first year of dam operation as they with time may lead to deterioration in the top of the dam (e.g., Mejia, 2021).

9.3 Deformed shape and potential cracking of the core

During the dam design analyses, the computed deformed shape and local radius of core curvature must be examined to evaluate whether the resulting computed magnitudes of extensional and shear strains are likely to cause cracking and leakage through the core. If that is the case, changes in design should be introduced or mitigating remedial measures to avoid detrimental crack formation. One such measure is to install horizontal reinforcement (e.g., armouring with bi-directional geogrids or precast reinforced lattice beams) in the top part of the dam to increase stability and reduce lateral and shear strains. The vertical spacing of such geogrids in the top part of the dam has in practice been in the range 1.0–2.5 m (see Sections 10.4 and 10.7). This way of strengthening the top of the dam against earthquake damage has been used in several large Chinese embankment dams and in some Russian dams. Analyses show that the reinforcement may reduce the displacements significantly, but to date, little field experience has been reported to corroborate the results of theoretical analyses. However, dynamic centrifuge testing of a model of a 130-m high ACED indicates that armouring with bi-directional geogrids in the top 30% of the dam may reduce the maximum tension stresses in the dam by about 35% and reduce the crest settlement by about 15% compared to the dam without armouring geogrids (Jiao et al., 2020). A similar reduction in tension and shear stresses and vertical settlements may be achieved by reducing the exterior dam slopes in the top part of the dam.

Another remedial measure to reduce the extent of cracking is to use a softer bitumen grade with a higher penetration value in the top part of the dam to tolerate larger extensional and shear strains before cracking occurs. Adding natural or steel or glass fibres in the bitumen increases the cracking resistance and the capacity to tolerate larger strains before cracking occurs. Nakamura et al. (2004) reported on testing with the use of a specially designed “Superflex-phalt”. By using additives in the bitumen, the HAC mix tolerated tensile strains during testing more than 30 times larger than when straight bitumen (B70) was used (see Section 3.10).

The shaking caused by the Whenchuan earthquake of magnitude 8 on the Richter scale in May 2008 caused the 156-m high Zipingpu concrete face rockfill dam (CFRD) to suffer significant damages with a settlement of 740 mm measured at the parapet wall and a horizontal residual displacement towards downstream of 230 mm, respectively (Liu et al., 2015). It is a well-known and documented case. The parapet wall was restrained by the upstream concrete facing, and boreholes revealed that there was a gap of about 200 mm under the baseplate of the parapet wall. Therefore, the maximum crest settlement of the dam was estimated to be about 1,000 mm. The magnitudes of measured settlements at the crest along the dam axis were roughly proportional to the dam height at the various cross-sections. The measured maximum settlement and horizontal displacement (towards downstream) recorded in the maximum dam cross-section at the dam centre line and 34 m below the crest were 810 and 100 mm, respectively.

The earthquake shaking caused severe damages to the concrete facing by deforming and separating it from the rockfill, by cracking and slab staggering and by concrete crushing in the vicinity of slab joints (Cruz et al., 2009). However, there were no signs of any critical dam slope sliding and instability. It should be noted that the reservoir level was only at about 1/3 of full supply level at the time of the earthquake shaking.

Based on the measured dam displacements and the dynamic response of the Zipingpu CFRD and by considering the more flexible and ductile behaviour of asphalt concrete compared to Portland cement concrete, one may infer that an asphalt core would have performed well without significant damage, if an ACRD of similar height as the Zipingpu CFRD had been built at that site.

A more flexible concrete facing with compressible infilling in the joints would probably have adjusted much better to the rockfill deformations that took place in the Zipingpu CFRD. The geomembrane-faced type of embankment dam rather than the CFRD will probably be selected more often at seismic sites in the future to take advantage of the membrane flexibility without undergoing cracking.

9.4 Measures to increase earthquake resistance

As demonstrated by extensive laboratory and field testing, asphalt concrete is a forgiving and ductile material that can tolerate large static and dynamic tensile and shear strains without cracking. However, during the design of an ACED, when the computed response for the design earthquake indicates that the core may undergo excessive deformations and critical cracking, mitigation measures are required to increase the dam's earthquake resistance. This concerns the top part of an ACED if it rests on a rigid rock foundation. For an embankment dam on a less rigid, compliant foundation, the dynamic amplifications are largest deeper down in the dam, often close to the dam-foundation interface. The following design and mitigation measures should be considered for ACEDs in very seismic regions:

- Use well-graded and well-compacted rockfill and gravel in the top part of the dam to reduce shear distortions, volumetric strains, and crest settlements. The effective stresses are low in the top part, and friction angles of crushed rock in the range 50°–60° may be achieved and used in the analyses. This reduces significantly the potential block (wedge) shear sliding computed by the Newmark type earthquake displacement analyses.
- Numerical deformation analyses must be performed for several plausible acceleration-time histories because the computed residual displacements and settlements are sensitive to the characteristics of the shaking, even when all the assumed acceleration-time histories are made compatible with the same design response spectrum.
- The top part of the embankment may be designed with more gentle exterior slopes than further down to increase stability in the top and reduce shear displacements and crown settlements.
- Use horizontal reinforcement, e.g., armouring with bi-directional geogrids or precast reinforced lattice beams to increase dam top stability and reduce settlements, displacements, and cracking (as described in Section 9.3).
- If the dam is founded on a deformable (compliant) foundation, use reinforcing tendons and anchors in the lower part of the upstream and downstream slopes where, according to the results of numerical analyses and field observations, local slope instability and significant shear distortions may occur. Dynamic analyses (linear and non-linear) may be performed by commercially available software to locate critical levels in the embankment cross-section.
- For an ACED on a rigid rock foundation, increase the core thickness in the top part of the dam to tolerate the dynamic amplifications that may occur and the deep cracks and larger shear displacements that otherwise could cause critical damage to the core.
- Use an asphalt concrete mix in the top part of the core with a softer bitumen grade (higher penetration value) and a higher bitumen content than deeper down in the core. This also

increases the rate of self-healing of any cracks that may have developed in the core. Introduce additives [e.g., styrene-butadiene-styrene (SBS) block polymer] to increase core flexibility and ductility and the ability to sustain large shear and bending strains without cracking. Reinforcing fibres may be added to the asphalt concrete mix to increase core cracking resistance (see Section 3.10).

- At the interface between the core and concrete plinth along the abutments, place a layer of sandy asphalt mastic (SAM) that may tolerate large interface shear strains that otherwise may cause cracking, leakage, and erosion.
- To ensure a watertight core-abutment interface or a core-concrete structure interface, water stops should be installed.
- For core-abutment or core-spillway interfaces steeper than about 60°, tilt (incline) the surface of the concrete plinth 15°–20° as shown in Figure 8.13. This promotes higher normal contact stresses and a tighter interface connection when the reservoir water pressure comes on.
- In case cracking and local leakage through the top part of the core should occur, Zone 2 on the downstream side of the core should be designed as a protective filter for the material in the upstream Zone 2 that should contain crack-stopper material. If cracks have developed and leakage has started, crack-stopper material should be placed (dumped) upstream of the leakage location. The graded filter on the downstream side of the core should be designed to arrest any erosion from upstream Zone 2 through cracks opened in the core. The gradation and mineralogy of the protective filter on the downstream side of the core should be such that it cannot "hold a crack" by developing cohesion or cementation.
- If cracks and leakage have developed through the core, local grouting in the upstream Zone 2 can be used to reduce leakage and potential erosion of the upstream Zone 2 material. At the same time, the crack self-healing process takes place due to the visco-elastic properties of HAC (see Chapter 5).

Chapter 10

Case studies of asphalt core embankment dams

10.1 Introduction

Field instrumentation, monitoring, and interpretation of observations have greatly advanced the development of the asphalt core dam concept and technology. Some of the early asphalt core embankment dams (ACEDs) that are not included among the detailed case studies described in this chapter are introduced as they have design and application features that have influenced the development and design of later ACEDs. Among them, the Eberlaste Dam in Austria, the Grosse Dhünn Dam in Germany, and the two High Island Dams in Hong Kong are briefly described.

10.1.1 Eberlaste ACED, Austria (1966–1968)

The Eberlaste asphalt core dam in Austria was built during the years 1966–1968 (ICOLD, 1992). The dam with a height of 28 m and a crest length of 480 m was built on top of a deep, compressible, and inhomogeneous alluvium deposit. The bedrock has a thin cover on the abutment slopes, and it dips steeply at about 60° towards the valley centreline. Bedrock was not encountered by a drill hole sunk 125 m deep in the middle of the valley filled with river-deposited gravel and sand. On the sides, the alluvial valley deposit is interlocked with talus material, and highly permeable material mixed with boulders predominates near the slopes. In the middle of the valley, the alluvial deposits are overlain by a silt and sand layer of about 20 m thick. There is no such layer on the sides near the slopes. The geological and geotechnical conditions suggested that big settlements and differential settlements would constitute the main challenges for the dam designers. Preliminary estimates suggested that about 1 m of settlement was to be expected, but as subsequent measurements showed, the settlements turned out to be more than twice that estimate.

A 0.6-m thick clay-cement cut-off wall was installed under the vertical asphalt core. It is 22 m deep in the middle of the valley and extends to a depth of 52 m in the permeable cones interspersed with boulders prevailing at the foot of the slopes. The thickness of the top part of the asphalt core is 0.4 m, and it is 0.5 m in the lower part. The aggregate recipe for the asphalt core at the time was 30% of coarse gravels (10–25 mm), 30% of fine gravels (3–10 mm), 32% of sands (0–3 mm), and 8% of limestone filler. The bitumen grade used was B300 and the bitumen content 8.0%. During construction, the dam foundation settled about 2.2 m in the middle of the valley, and during the 2 years following the end of dam construction, a secondary central settlement of 20 cm was measured. Using the recorded dam foundation settlement profile across the valley as the input, Høeg (1995) carried out a finite element analysis to back-calculate the stresses, strains, and shear distortions that the core must have undergone.

DOI: 10.1201/9781315177786-10

No noticeable leakage has been recorded through the asphalt core in the Eberlaste Dam since construction of the dam about 60 years ago. This must mean that the core has been able to adjust to the large differential settlements and shear and extensional strains without cracking, leading to an increase in asphalt core permeability. The designers had the foresight to specify the use of an especially soft grade bitumen (grade B300) and a high bitumen content (8.0%) in the asphalt concrete to accommodate the anticipated large differential settlements and shear distortions without cracking. The bitumen grade used had a much higher penetration value and the asphalt concrete a higher bitumen content than applied in any of the hydraulic structures built up to that time. The average penetration value of the bitumen was 269 (0.1 mm) of 138 control tests taken during construction. There was no previous field experience with using so soft bitumen in the dam barrier. Furthermore, the deep clay-cement cut-off wall installed in the alluvial foundation was made somewhat compressible in an attempt to match the settlements in the alluvium on both sides of the cut-off wall (see Section 8.6). In several ways, the design and construction of the Eberlaste Dam was a pioneering project, and the success established a precedence for later ACED designs on compressible, inhomogeneous, and pervious soil foundations.

10.1.2 High Island ACEDs, Hong Kong (1973–1978)

The two High Island rockfill dams in Hong Kong built in 1973–1978 have a unique design for a special purpose and were the highest asphalt core dams for many years (Schönian, 1999). In the 3.5 km long Guanmen Strait between the Chinese mainland and the High Island, the East Dam and West Dam were built at each end of the Strait to close a reservoir with an effective storage capacity of $2.74 \times 10^8 m^3$ at storage level el. 61.0 m.asl. and the largest water supply reservoir for Hong Kong with the largest supply of $56 \times 10^4 m^3$/day.

The heights of East Dam and West Dam are 107 and 102.5 m, respectively, with crest lengths of 490 and 760 m. The two dams are ponded on both sides. One side of the dam impounds drinking water and the other side sea water. The difference between reservoir water level and sea water level amounts to around 60 m. It was therefore necessary to control leakage water on both sides, on the sweet water side and on the saltwater side. The two dams were designed with double asphalt cores, and the secondary asphalt core wall is situated 2.8 m seawards of the main asphalt core wall, to keep off the sea water. The top of the secondary asphalt core wall is around 10 m higher than the mean sea level to protect even against the highest anticipated typhoon floods.

The asphalt core walls are founded on a 12 m wide concrete plinth, wide enough to give space for 5 parallel rows of injected grout curtains. The plinth slabs (segments) are maximum 10 m long, and their contraction joints are filled with sandy asphalt mastic.

The core wall foundation with an inspection gallery is founded on sound rock, 36 m below sea level for the West Dam and 44 m below sea level for the East Dam. The inspection gallery is positioned 15 m above mean sea level inside the seaside shoulder of the dam. Both core walls are connected to the gallery by means of asphalt concrete blankets. In case of a defect in the main core wall below the upper blanket, leakage water would enter the gallery by one of the lower drainpipes between the two blankets. Should a leak occur in the upper portion of the main core wall, one of the upper drainage pipes would convey the leakage water into the gallery. To easier be able to locate eventual defects, the space between the core walls below the gallery is separated by cross walls at 60 m apart. The main asphalt core wall is 1.2 m wide in the lower section and 0.8 m in the upper section. The secondary core wall is 0.6 m wide.

In 1993, increased leakage was discovered at the plinth of the right abutment of the West Dam. The leakage location was not through the asphalt core wall but 20 m below the dam crest and beneath the asphalt core base. After grouting, the foundation leakage was successfully reduced from 6.3 L/second to less than 0.1 L/second.

10.1.3 Grosse Dhünn ACED, Germany (1983–1984)

The Grosse Dhünn asphalt core dam seated on a rock foundation with a dam height of 63 m was completed in 1984 (Schönian, 1999). The asphalt core was designed vertical in the lower 2/3, but it was inclined (kinked) in the upper part to increase the contact stresses between the upstream transition zone and the core. This was intended to reduce displacements in the top of the upstream dam slope and prevent detachment between the transition zone and the core. A special floating shaft was installed downstream of the core to measure displacements of the core and in the transition zone along the interface between the core and the transition zone. The innovative floating shaft and instrumentation arrangement, used for research and developmental purposes, proved very successful. Based on the observed behaviour and interpretation of results from the Grosse Dhünn ACED and several other ACEDs, Schönian (1999) concluded that (quote) "there are no indications to put in question the application of asphaltic concrete cores for future dams much higher than 100 m".

Key references

ICOLD. 1992. *Bituminous Cores for Fill Dams - State of the Art.* Bulletin 84. International Commission on Large Dams (ICOLD), Paris.

Schönian, E. 1999. *The Shell Bitumen Hydraulic Engineering Handbook.* Thomas Telford, London.

Several case studies of ACEDs are presented in the following pages. Some of the older ACEDs presented do not satisfy all the requirements and technical specifications described in this book because the specifications did not exist at the time of dam design and construction.

10.2 Finstertal Dam, Austria

The Finstertal Dam was built in 1977–1980 to create a reservoir for seasonal storage at high altitude for the Sellrain-Silz scheme in the Austrian Alps about 30 km west of Innsbruck (Schwab and Pircher, 1985; Pircher and Schwab, 1988; ICOLD, 2018).

10.2.1 Dam site conditions

The severe climatic conditions at an altitude of more than 2,300 m.asl. permitted only about 100 days a year for embankment construction. The special topography of the site (Figure 10.1) causes the dam to straddle the rock sill, which bars the glacial trough of the reservoir basin and where a former natural lake had its outlet. An optimum position with respect to embankment volume and stability was achieved with a double curved layout of the dam axis, i.e., with a slight upstream curvature (convex) of the main centre part and a short reverse curve (concave) of the right wing related to the complex rock foundation geometry. As a result, the asphalt core seated on the ridge has a maximum height of 96 m, while the dam with a crest length of 652 m has a

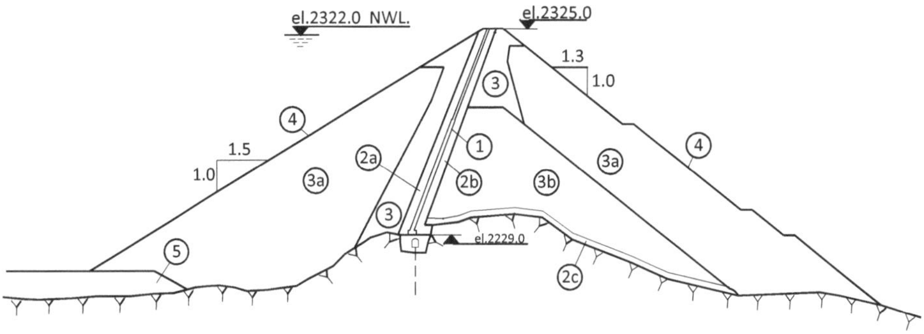

Zone	Material	Aggregate d_{max} (mm)
(1) Impervious core	Asphalt concrete	16
(2a) Upstream transition zone	Screened moraine	100
(2b) Downstream transition zone	Crushed granodiorite rock	100
(2c) Drainage zone	Quarried rock	700
(3) Shoulder	Quarried rock	700
(3a) Shoulder	Quarried rock	700
(3b) Shoulder	Moraine	700
(4) Riprap	Blocks	500–1,000 (length of stone)
(5) Remaining overburden	Moraine	

Figure 10.1 Cross-section of the Finstertal Dam, Austria (ICOLD, 2018).

maximum height of 155 m measured from the downstream toe. The dam crest width is 9 m, and the dam slopes are steep with an upstream slope of 1V:1.5H and 1V:1.3H for the downstream slope.

10.2.2 Dam design and zoning

The asphalt core was designed to incline over the full height with a slope of 1V:0.4H, and the asphalt core width was changed in three steps with 0.70, 0.60, and 0.50 m from the bottom to the upper part. The volume of the asphalt core is 24,000 m^3. The asphalt core is composed of 0–16 mm aggregates with a filler content of 12%. The filler consists of 8% limestone powder plus 2%–3% reclaimed filler from the crushing of aggregates. The bitumen content (B65) is 6.3%. The widths of the upstream and downstream transition zones are 2 and 3 m, respectively. Figure 10.1 shows the cross-section of the Finstertal Dam.

The dam fill materials consist of solid and sound granodiorite and stiff moraine. To increase contact sliding resistance between the rockfill and the smooth foundation surface that was caused by glacial erosion, key slots with a mean depth of 0.7 m were excavated in a foundation surface area of 20,000 m^2. The volume of the dam is 4.4 million m^3. The dam creates a reservoir of a total storage capacity of 60 million m^3. Granodiorite was used for the upstream dam shoulder (shell), and granodiorite and moraine were used for the downstream dam shoulder (shell). The rockfill was placed with about 0.75 m layer thickness, water sluiced, and compacted by vibratory rollers. The compacted void content (porosity) was 21%–24% for the upstream dam shoulder.

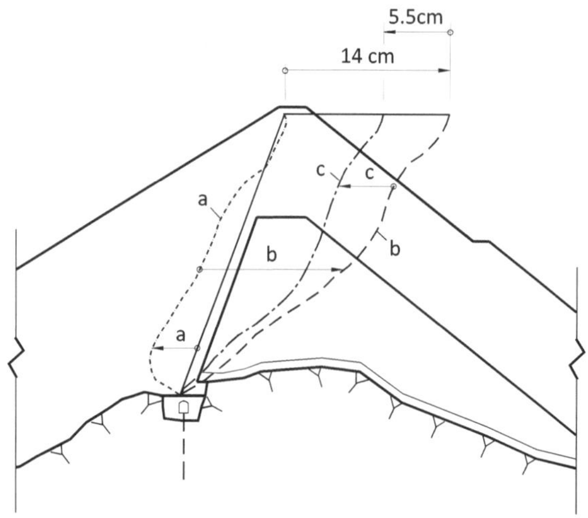

Figure 10.2 Horizontal displacements of the asphalt core of the Finstertal Dam. (a) At end of construction, 1980; (b) after first impounding to max. storage level, 1981; and (c) first complete drawdown, 1984 (Schwab and Pircher, 1985).

10.2.3 Dam instrumentation and monitoring

Figure 10.2 shows the horizontal displacements of the asphalt core of the Finstertal Dam at the end of construction in 1980; at first impounding to max. storage level in 1981; and at first complete drawdown in 1984. As may be noticed, a significant rebound was measured when the reservoir was lowered.

The downstream dam shell is very stiff, and the overall deformations of the dam are very small, considering the dam height. The back-calculated compression modulus was found to be very high due to the heavy compaction and relatively thin rockfill layers used during construction. At the end of dam construction in 1980, the core had moved in the upstream direction with a maximum value of 4.0 cm in the lower part of the core wall due to the core seating at the upstream side of the rock ridge. When the reservoir was first impounded to the maximum storage level in 1981 and after first complete drawdown in 1984, the top of the core wall had moved in the downstream direction with a value of 14.0 cm and then rebounded 5.5 cm at first complete drawdown (Figure 10.2).

In 2008, 40 years after end of construction, the maximum vertical settlement of the dam crest was 38 cm and the maximum horizontal displacement 19 cm. The seepage from the whole asphalt core area of 37,000 m^2 was only 9 L/second after the first impounding in 1981, and it reduced to 5 L/second in 2008 and to less than 3 L/second in 2009 due to the migration of fine particles and clogging of seepage paths.

Key references

ICOLD. 1992. *Bituminous Cores for Fill Dams - State of the Art.* Bulletin 84. International Commission on Large Dams (ICOLD), Paris.

Pircher, W., and Schwab, P. 1988. Design, construction and behaviour of the asphaltic concrete core wall of the Finstertal Dam. In *Proceedings of the 16th ICOLD Congress*, R49-Q61, San Francisco, CA; pp 901–924.

Schwab, H., and Pircher, W. 1985. Structural behaviour of a high rockfill dam, comprehensive interpretation of measurements, and conclusions on stress-strain relationships. In *Proceedings of 15th International Congress on Large Dams (ICOLD)*, Lausanne, Switzerland; pp. 1273–1297.
Schönian, E. 1999. *The Shell Bitumen Hydraulic Engineering Handbook*. Thomas Telford, London.

10.3 Storglomvatn Dam, Norway

The Storglomvatn Dam is the bigger of two asphalt core dams in Statkraft's 600 MW hydropower development by the Arctic Circle near the Svartisen glacier in northern Norway (Karlsen et al., 1995). At the design stage, the options of using an earth core (till) or a concrete facing were compared with the design of an asphalt core dam. The asphalt core was chosen based on economics and the very successful construction and performance of several previous asphalt core dams at the time in Norway, particularly the 90-m high Storvatn dam (Arnevik et al., 1988).

10.3.1 Dam site conditions

The left abutment mainly consists of meta-sandstone, while the right is of folded limestone with intercalations of mica schist/mica quartz/mica gneiss and hornblende schist. The limestone (mainly marble) is karstic with dissolution channels and cavities and represented a challenge to the dam designer. Sizeable cavities and channels in the limestone were found on the rock surface, and the concern and question were how deep the karstification reached. Several boreholes, vertical and inclined, were drilled down to 100 m depth, and seismic tomography and georadar were used to explore the foundation. Not much karst was detected by these methods, but it was decided to construct a gallery about 20 m under the surface along the right abutment. The purpose of the gallery, 5 m in diameter, was to explore the foundation in more detail, facilitate grouting, and give the possibility of post-construction grouting if that proved to be required. The gallery revealed significant karstic features, and that information was used in setting up the grouting program. Due to the surface karstification, deep core trenches had to be excavated for the concrete plinth under the asphalt core along the right abutment.

10.3.2 Dam design

The Storglomvatn Dam, 128 m of max. height and volume of $5.3 \times 10^6 m^3$, is sited on a stiff rock foundation. The crest length is 825 m with a slight upstream (convex) curvature of radius 1,500 m. The vertical axis of the core was located 1.5 m upstream of the dam axis. Figure 10.3 shows the zoning of the dam fill material. The dam height at that cross-section is 123 m. The upstream slope is 1V:1.5H, the downstream slope between the berms is 1V:1.4H, and the four berms are between 2.5 and 2.9 m wide. The width of the dam crown is 7 m. A special design consideration for the dam was that the rockfill available locally was much inferior to that used for previous Norwegian asphalt-core dams and that was accounted for in the dam deformation analyses performed and in the final design.

10.3.3 Dam construction and performance

Table 10.1 gives the zoning and compaction specifications for the various zones. The natural gravel used for the asphalt concrete aggregates and for the transition Zone 2 had varying petrographic and mineralogical origin. The marble rockfill from the Storglomvatn quarry within the reservoir was of relatively poor quality, and significant breakage occurred during handling in the quarry and during placement and compaction on the dam. The unconfined compressive strength

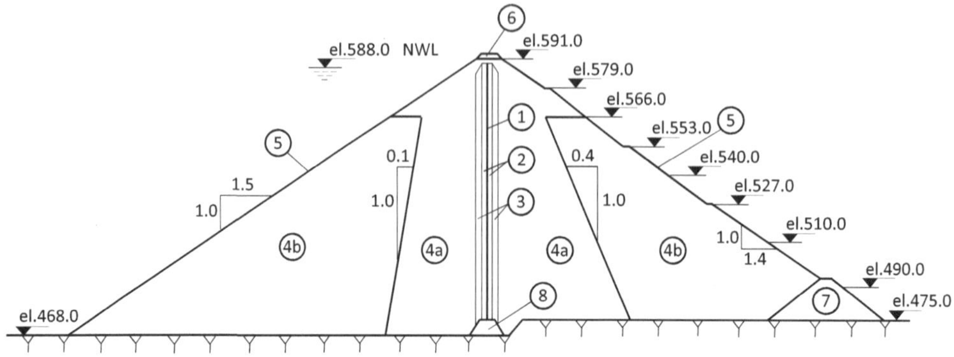

(1) Asphalt core; (2) transition zone of natural gravel; (3) transition zone of crushed rock; (4a) well-compacted quarried rock; (4b) shoulder of quarried rock; (5) riprap; (6) crown cap protection (blocks > 1 m³); (7) toe drain (blocks > 0.5 m³); and (8) concrete plinth.

Figure 10.3 Design and zoning of cross-section of the Storglomvatn Dam.

Table 10.1 Zoning and compaction specifications for the Storglomvatn Dam (Høeg et al., 2007)

Zone	*Material*	*Compacted layer thickness (m)*	*Compaction*	
			Vibratory roller weight (tonne)	*Number of passes*
(1)Asphalt core	Bitumen 180 aggregate (0–18 mm) filler (<0.075 mm)	0.2	Trials on site approx. 0.5	Until air porosity <3%
(2)Fine transition	Gravel (0–60 mm)	0.2	Approx. 1.5	3–6
(3)Coarse transition	Crushed rock (0–200 mm)	0.4	15+water sluicing	4
(4a) Shoulder (Shell)	Quarried rock (0–700 mm)	0.8	15+water sluicing	8
(4b) Shoulder (Shell)	Quarried rock (0–1,000 mm)	1.6	15	6

(UCS) of the marble in dry condition was in the range of 40–60 MPa but only half of that in wet condition. Therefore, water sluicing (about 500 L/m³) was specified for Zones 3 and 4a to reduce the rockfill porosity during construction and to reduce the post-construction settlements when the upstream rockfill became submerged. The downstream fill also becomes wet with time and settles because of precipitation and melting snow.

The width of the concrete plinth under the base of the asphalt core is 6 m at the bottom of the valley decreasing to 4 m higher up along the abutments. The asphalt core is 0.9 m wide at the bottom decreasing to 0.5 m at the top. The top 51 m has a constant width of 0.5 m. Transition Zone 2 is 1.3 m wide at the bottom increasing to 1.5 m in the upper part. The total width of the core and the transition Zone 2 is held constant and equal to 3.5 m over the entire height, to facilitate that the core and the transition Zone 2 can be placed simultaneously in one operation by the paver. The width of transition Zone 3 is 4.0 m constant over the whole dam height. As the construction and post-construction deformations for the Storglomvatn Dam

Figure 10.4 Storglomvatn Dam close to completion in 1997. (Courtesy of Statkraft.)

were expected to be larger than in previous Norwegian asphalt core dams and as the dam is in a region with some earthquake activity, the designer specified an asphalt mix with a softer grade bitumen (B180) and somewhat higher bitumen content (6.7%) than in the earlier ACRDs built in Norway.

The climate is cold and foggy with high precipitation, and the dam was constructed in four construction seasons from late May to middle of October during the years 1993–1997. The construction season was short and during one year the site was not accessible until mid-June, due to the heavy snowfall in the area. Figure 10.4 shows a photo of the dam towards the end of construction in 1997. The reservoir was simultaneously slowly impounded by the melting snow and water from the glacier located upstream.

10.3.4 Measured dam displacements

The designer did not require the installation of displacement sensors or slope indicators in the interior of the dam as much knowledge and confidence had been gained from the successful construction and performance of several previous Norwegian asphalt core dams, particularly the well-instrumented and analysed 90-m high Storvatn Dam (Arnevik et al., 1988; Adikari et al., 1988). However, surface measuring bolts (57 on the downstream slope and 36 on the upstream) were installed during construction as soon as the specified fill elevation was reached. Geodetic surveys were used to measure vertical settlements and radial and tangential displacements (in the direction along the axis of the dam). Bolts were also installed in the top of the core at 8 points along the dam axis, 3 m under the level of the dam crown.

Vertical and horizontal displacements of the crown, the top of the core, and of the downstream slope were monitored during construction, impounding, and dam operation. Impounding of the reservoir took place at the same time as the dam was built, making early power production possible. Table 10.2 shows the observed settlements and horizontal displacements of the downstream slope at the end of dam construction and impounding for the Storglomvatn Dam.

Table 10.2 Observed displacements on downstream slope at the end of dam construction and impounding for the Storglomvatn Dam (see Figure 10.3)

El. (m.asl.)	475 (Rock found.)	490	510	527	540	553	566	579	591 (Dam crest)
Settlement (m)	0	0.06	0.34	0.44	0.435	0.32	0.14	0.02	0
Hor. displacement (m)	0	0.30	0.48	0.32	0.26	0.20	0.07	0.04	0

The observed surface displacements were used to back-analyse the internal behaviour of the Storglomvatn Dam (Wang, 2008). The conclusions were that the combination of the transition zones and the stiff rockfill Zone 4a ensured that the asphalt core experienced low bending and shear stresses and a maximum vertical compressive strain of only around 1.0%, which is well below the strain level when fissuring and cracking in the core might occur (see Figure 4.1). The rockfill and gravel embankment govern the deformation of the core. As shown by parameter studies, the deformation modulus of the core itself has virtually no influence on the lateral core and embankment deformations.

After 9 years of operation (in 2006), the vertical and radial displacements at the three locations, i.e., the crown on the upstream side (el. 591.0 m), the top of the core (el. 588.0 m), and the crown on the downstream side, were 0.38/0.08/0.28 m and 0.08/0.13/0.21 m, respectively (Høeg et al., 2007). The crown settlement on the upstream side is significantly larger than the crown settlement on the downstream side. This is normal behaviour as the upstream rockfill settles due to submergence and wetting, and the seasonal raising and lowering of the reservoir (slow cyclic loading) creates additional settlements in the upstream shoulder. At Storglomvatn Dam, the reservoir filling lagged the rockfill construction level. Therefore, the extra settlements due to submergence of the top part of the dam appeared only after dam construction was completed and the settlement bolts were installed. The fact that the reservoir filling turned out to be slower than anticipated led the designer during the last construction season to increase the maximum camber for the dam by 0.2 m, from 1.0 to 1.2 m.

The downstream side of the crown moved horizontally significantly more in the downstream direction than the upstream side of the crown causing an extensional strain across the crown. Longitudinal cracks often develop in the crown of embankment dams. The cracks are not a safety concern as they are longitudinal (not transversal) and shallow.

10.3.5 Seepage measurements

Leakage was of concern for the Storglomvatn Dam due to the karstic foundation, and the strike of the bedding planes was approximately normal to the dam axis. Seepage measurements were taken during reservoir filling and for 9 years of operation. The recorded seepage data show fluctuations each year due to precipitation and water from snow melting, some of which enters the measuring weir. When this is accounted for, there is a decrease in measured seepage/leakage from year to year, and the measured leakage is very small (less than 7 L/second at NWL). It is decreasing with time due to the crack-stopper effect by silt and fine sand migrating into joints and cracks under the dam. The grouting of the karstic foundation seems to have been very successful, and there has been no need for post-construction grouting. It should be noted that the installed seepage weirs will not collect all flow deep in the foundation, but there has been no observation of increased flow in the small brooks downstream of the dam after the reservoir was filled.

Key references

Høeg, K. 1993. *Asphaltic Concrete Cores for Embankment Dams*. Stikka Press, Oslo.
Høeg, K., Valstad, T., Kjærnsli, B., and Ruud, A. M. 2007. Asphalt core embankment dams: recent case studies and research, *Hydropower and Dams*, 13(5):112–119.
ICOLD. 2018. *Asphalt Concrete Cores for Embankment Dams*. Bulletin 179. International Commission on Large Dams (ICOLD), Paris.
Wang, W. 2008. Research on the suitability of asphalt concrete as water barrier in dams and dikes. PhD thesis, University of Oslo, Department of Geosciences, Norway.

10.4 Yele Dam, China

The Yele project, on the very upper reaches of the Nanya River in the southwest of Sichuan Province, is the most upstream of six projects in a cascade development for electricity generation. The river is 49.5 km long with a hydraulic head drop of 1,714 m and exploitable hydropower of 700 MW. The topography of the Yele basin is well suited for establishing a reservoir with a catchment area of 323 km^2.

10.4.1 Dam site and foundation conditions

At Yele, the winter season is 6–7 months long, and the rainy season is from May to October. Annually, there are about 215 rainy days with a mean rainfall of 1,830 mm and an air relative humidity of 86%. The annual mean temperature is 7°C, ranging from –20°C to +28°C.

Figure 10.5 shows the geological conditions at the dam site, and Wang et al. (2010a) and Wang and Høeg (2010) described the complex conditions in more detail.

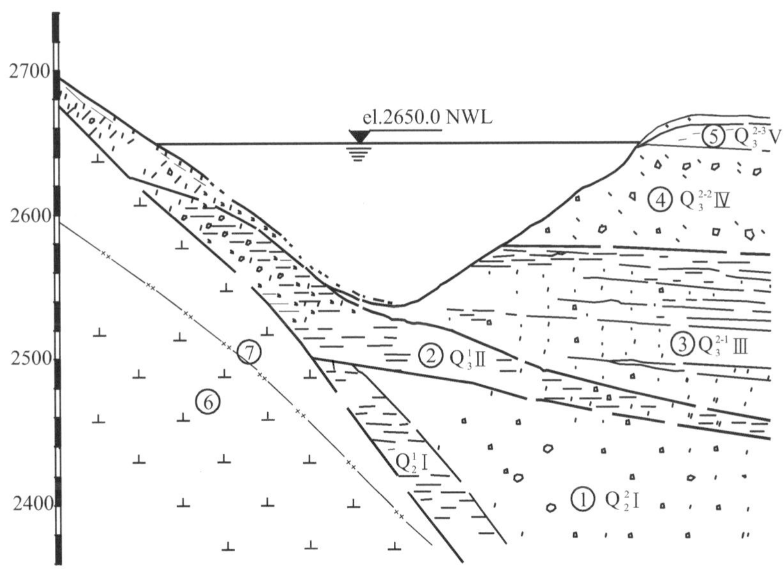

(1) gravel with silt and sand layers; (2) stiff, over-consolidated cohesive soils with stones; (3) gravel with layers of loam; (4) gravel; (5) sandy soil with loam and carbonized plant fragments; (6) quartz diorite bedrock; and (7) crevice/lineament.

Figure 10.5 Geological cross-section of the Yele Dam foundation and abutments.

10.4.2 Dam type selection and design

For the difficult geological foundation conditions with an irregular and compressible overburden and with the high regional seismicity, only an embankment-type dam was considered feasible. Three options were examined for the impervious barrier in a rockfill dam: earth core rockfill dam (ECRD), concrete face rockfill dam (CFRD), and asphalt core rockfill dam (ACRD). To select among these options, an emphasis was placed on costs, sensitivity to severe weather conditions during construction, earthquake resistance, and compatibility with the geological conditions that may cause significant differential settlements across the valley.

In the rainy and cold Yele area at high altitude (crest elevation at el. 2654.5 m.asl.), the water content of the earth core material in the local borrow was 10 percentage points higher than that required for optimum compaction. It would be very difficult and time-consuming to reduce the water content, and core placement would have to be stopped during the frequent rainy periods. For the CFRD option, the concrete slabs would be placed after the upstream slope was completed, and it was considered difficult to protect the upstream slope from damage by sudden heavy rainstorms during the dam construction period. Furthermore, the impounding could not commence until the dam with a concrete face was completed. This was a disadvantage at the Yele site, where the reservoir would take a long time to fill as the rate of annual river flow is low. The core for an ACRD may be constructed during periods of rain and cold weather. During heavy rains, the asphalt mix is stored in hot silos. When the heavy rain stops, the asphalt core construction can be restarted immediately after cleaning and heating the asphalt surface, without the long delay associated with earth core construction in rainy weather.

The CFRD requires a longer concrete plinth than the ACRD, and if large differential settlements occur, leakage may develop in the joints between the slabs and in perimeter joints due to rupture of water stops. At the Yele site, such settlements could be caused by the non-uniform geological foundation conditions and severe earthquakes, which may cause large in-plane stresses in the concrete face. On the contrary, the ACRD with a central asphalt core, if properly designed, is sufficiently flexible and ductile to be able to accommodate the anticipated differential settlements without cracking.

For the site and environmental conditions at Yele, the ECRD was estimated to cost approximately 10% more than the ACRD, while the CFRD was estimated to cost around 10% less than the ACRD. Among the three options, all aspects considered, the ACRD was selected as the most suitable.

Figure 10.6 shows the maximum cross-section of the Yele Dam. The asphalt core in that cross-section is 120 m high and the total dam height is 124.5 m. Figure 10.7 shows a plan view of the dam with the locations of cross-sections A–G. Figure 10.8 shows a view of the completed dam. As designed, the Yele asphalt-core rockfill dam has a crest length of 411 m with a 300 m long seepage cut-off wall extension along the top of the right bank. Due to the very high seismicity of the region, the dam was designed with gentle slopes of 1V:2H upstream and 1V:2.2H downstream, and a wide crest (14 m). In addition, as an earthquake-resistant measure, geo-grids were placed horizontally to reinforce the top 30 m of the dam (from el. 2624.5 m.asl. to the dam crest at el. 2654.5 m.asl.) (see Section 9.3). In the upper 20 m, the vertical height difference between the geo-grids is 1 m, and within the lower 10 m, it is 2 m. Furthermore, to strengthen the lower part of the upstream dam slope against large deformations and potential sliding during severe earthquake loading, 40 m long geo-grids were placed horizontally from the upstream dam face between el. 2,594 m.asl. and el. 2,603 m.asl. The vertical height difference between the 6 (six) geo-grid layers is 1.5 m (Wang et al., 2010a).

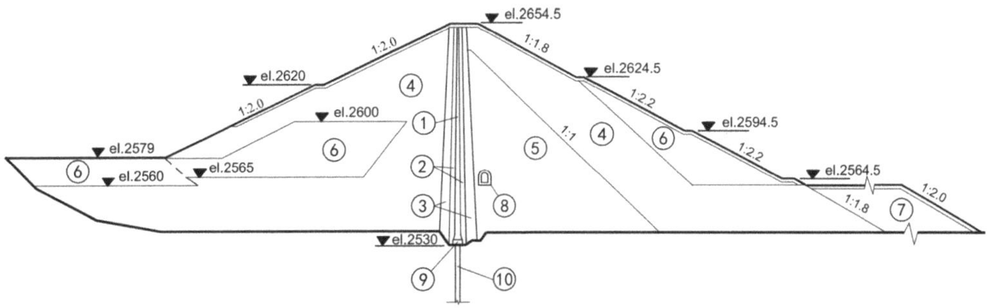

① Asphalt core, 1.2 m width at the bottom, gradually reduced to 0.6 m at the top; ② fine transition zone, 1.5 m width; ③ coarse transition zone, 4.0 m width at the bottom and 2.0 m at the top; ④ rockfill (I); ⑤ rockfill (II); ⑥ natural gravel and rockfill (III); ⑦ toe berm (22 m in thickness and 215 m in length); ⑧ observation gallery for field instrumentation (see Figure 8.12); ⑨ concrete plinth (see Figures 8.4 and 8.5); and ⑩ concrete cut-off wall (see Figure 10.9).

Figure 10.6 Cross-section of the Yele Dam (this is section D shown in Figure 10.7).

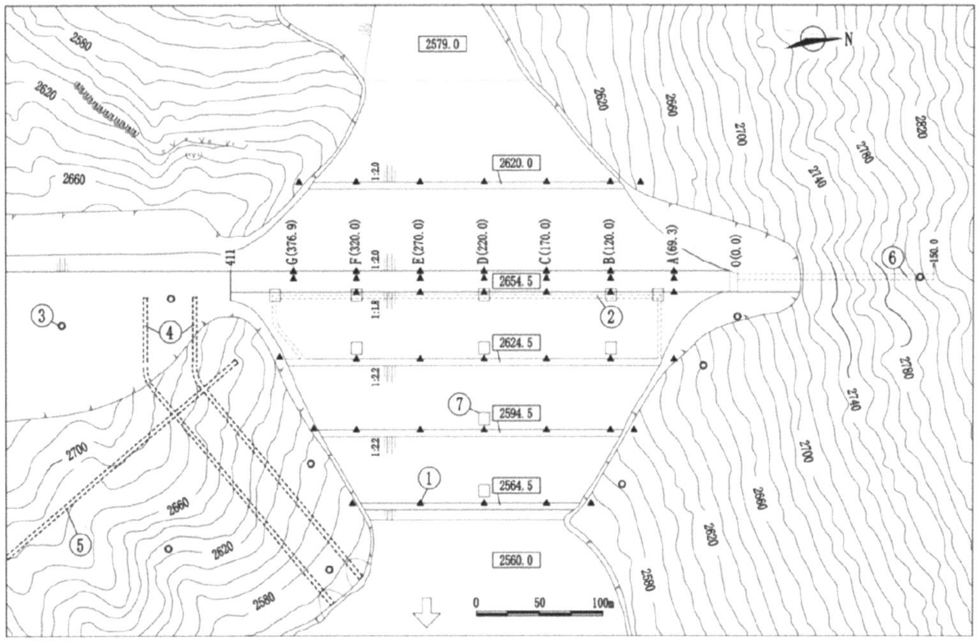

① Displacement bolts; ② observation gallery; ③ piezometers; ④ access galleries; ⑤ drainage gallery; ⑥ grouting gallery; and ⑦ observation huts.

Figure 10.7 Plan view of the Yele Dam and location of monitoring instruments.

The vertical core wall, located 3.7 m upstream of the dam centreline, was designed to be 1.20 m wide at the bottom and decreasing gradually to 0.60 m at the top (el. 2,653 m.asl.). The base of the core is flared out against the plinth to a width of 2.40 m at the core-plinth interface. Similarly, the core is flared out against the plinth at the abutments to twice the core width at that elevation.

Figure 10.8 View of the Yele Dam in operation in 2017.

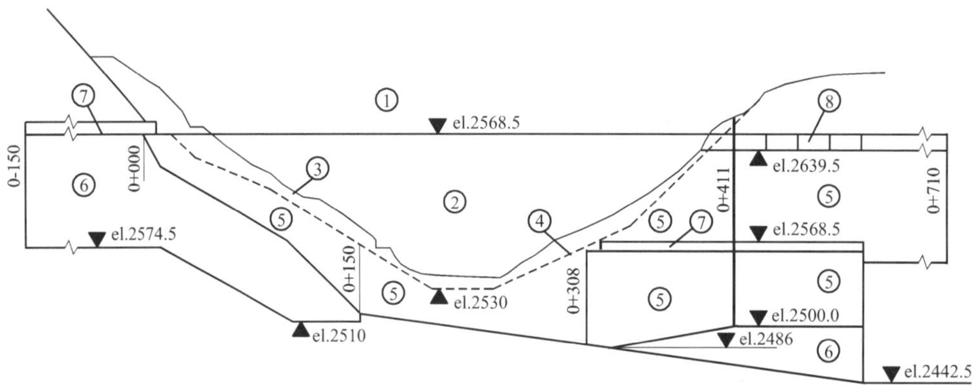

① Crest; ② asphalt core; ③ ground surface; ④ excavation line; ⑤ concrete cut-off wall; ⑥ grout curtain; ⑦ construction gallery used for construction of grout curtain and concrete cut-off wall; and ⑧ concrete cut-off wall extension built in open excavation.

Figure 10.9 Water barriers in the Yele Dam foundation.

The design of the critical asphalt core – plinth connection at the top of the concrete cut-off wall is shown in Figures 8.4 and 8.5.

10.4.3 Design and construction of impervious barriers in the foundation

Figure 10.9 shows the complex system of impervious barriers installed in the foundation to reduce and control the under-seepage, and Wang et al. (2010a) described the design and construction in more detail.

10.4.4 *Yele Dam construction and performance*

The Yele asphalt-core construction was started in November 2003 and was completed in November 2005. Zoning, compaction specifications, and quality control for the Yele Dam are shown in Table 10.3. Figure 10.10 shows the progress of dam construction and impounding until October 2007. Figure 10.11 shows the dam under construction and simultaneous impounding of the reservoir.

The Yele Dam embankment construction started in April 2001, and in December 2004, the dam reached el. 2,603 m.asl., which is 51.5 m below the crest of the dam (el. 2,654.5 m.asl.). In January 2005, a special impounding test was started as a quality control of the construction work done and whether there was any leakage at the base of the asphalt core. At that time, the water level was at el. 2,552 m.asl. behind the upstream cofferdam, as shown in Figure 10.10. After 1 month of impounding, the water level was raised by 35.5 m to el. 2,587.5 m.asl. Subsequently, the reservoir was lowered to its original level. Observations were made of the pore-water pressures on the downstream side of the core, of the deformations of the core, and of the strains in the concrete plinth and cut-off wall during the raising and lowering of the reservoir while the embankment height was kept constant at el. 2,603 m.asl. The pore-water pressures on the downstream side of the core wall were measured to be close to zero during the impounding test, and the deformations of the asphalt core and the strains in the concrete plinth and cut-off wall were measured to be very small. On 9 March 2005, the reservoir water level was raised

Table 10.3 Zoning, compaction specifications, and quality control for the Yele Dam

Zone (see Figure 6.6)	*Material*	*Layer thickness (m)*	*Compaction by vibratory roller*	*Quality control*
Asphalt core	Bitumen B70 Aggregate (0–20 mm)	0.20–0.26	8 passes, 1.5 tonnes	Air porosity <3.0%
Fine transition zone 1.3–1.6 m	Gravel (0–80 mm, grain size <5 mm passing 20%–40%; grain size <0.075 mm passing ≦10%	0.20–0.26	4 passes, 2 tonnes	Dry density ≥ 20.6 kN/m³, porosity ≦ 20%
Coarse transition zone 2.0–4.0 m	Gravel (0–150 mm, grain size <5 mm passing 10%–20%; grain size <0.075 mm passing ≦3%	0.20–0.26	4 passes, 2 tonnes	Porosity ≦ 22%
Shoulder (I)	Quarried rock (0–800 mm)	1.0–1.2	8 passes, 20 tonnes	Dry density≥ 21.9 kN/m³, porosity ≦24%
Shoulder (II)	Quarried rock (0–800 mm)	1.0–1.2	8 passes, 20 tonnes	Dry density≥ 22.5 kN/m³, porosity≦22%
Shoulder (III)	Natural gravel or quarried rock (0–800 mm)	1.0–1.2	8 passes, 20 tonnes	Dry density≥ 21.1 kN/m³, porosity ≦22% (upstream), 24% (downstream)
Toe berm	Rock debris (0–1,200 mm)	1.0–1.2	8 passes, 20 tonnes	Dry density≥ 19.6 kN/m³

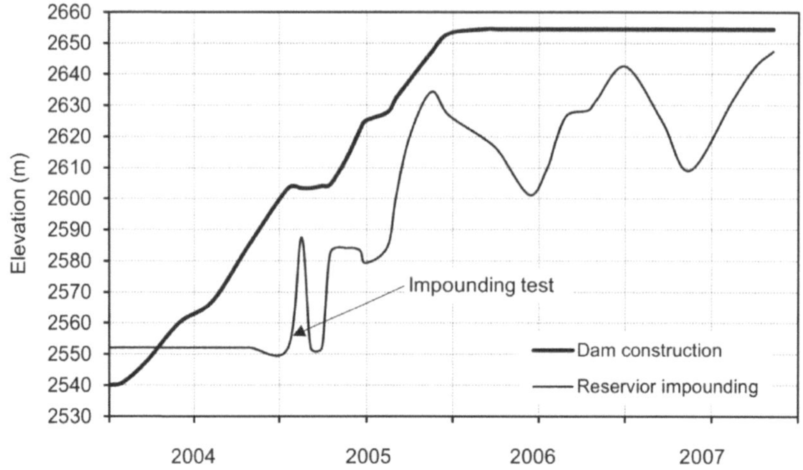

Figure 10.10 Progress of dam construction and reservoir impounding. (Note: the reservoir was held back by the upstream cofferdam before the special impounding test started at el. 2,552 m.asl.)

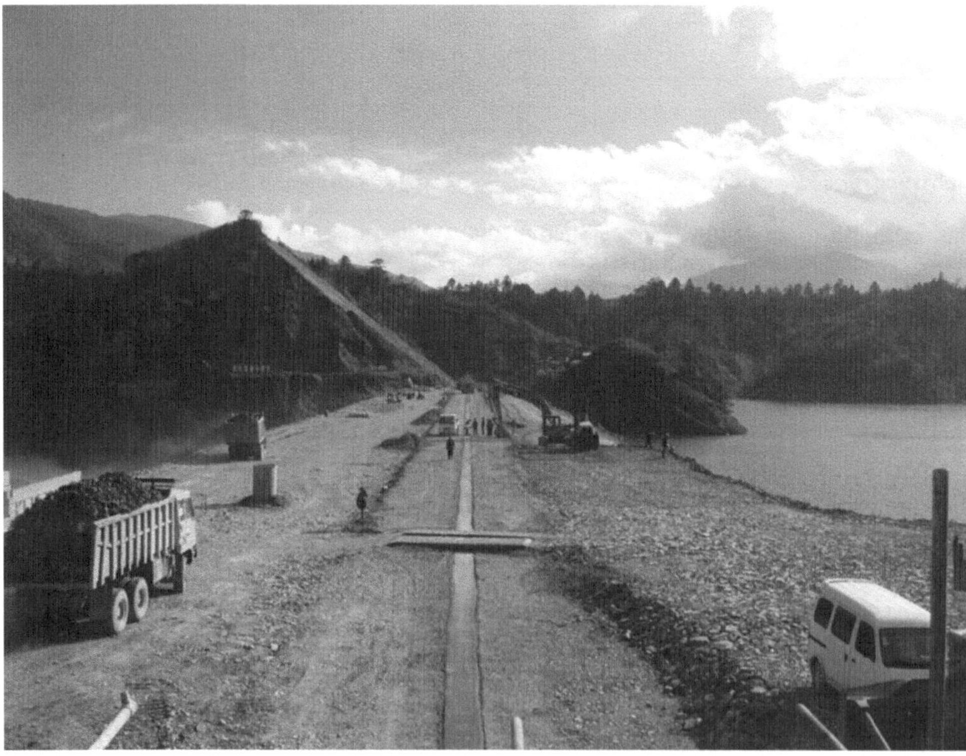

Figure 10.11 Yele Dam under construction and simultaneous impounding of the reservoir.

again and reached el. 2,634 m.asl. on 26 October 2005. The embankment reached the crest elevation 2654.5 m.asl. in December 2005.

During construction, impounding, and operation, the dam was monitored by means of a comprehensive instrumentation system.

10.4.5 Measured dam surface displacements

Post-construction displacement observations at the top of the vertical core and at the upstream and downstream sides of the crown started on 21 June 2006. Unfortunately, this is about 6 months after the end of dam construction, so the post-construction displacements during the first months are not included. The maximum post-construction settlement is measured to be 45 mm at the upstream side of the crown. The settlement may have been almost twice that if one were to include the estimated post-construction settlements during the first 0.5 year after the end of construction.

Until January 2009, the dam had experienced three cycles of water level rising and lowering, and the dam had been subjected to 10 earthquakes of various magnitudes that gave relatively small vibrations at the Yele site. The maximum post-construction settlement was measured at the upstream side of the crest at section D, and the value was about 140 mm at that time.

When the Wenchuan earthquake struck in May 2008, the Yele reservoir level was near the minimum operating level el. 2,600 m.asl. The maximum additional crest settlement during that earthquake was about 15 mm at section D. The Yele Dam site is located 258 km from the epicentre of the Wenchuan earthquake (magnitude 8.0), and the intensity (Chinese scale) at the dam site was less than VI. According to the monitoring of the nine strong-motion seismographs installed on and inside the Yele Dam, the calculated maximum settlement and horizontal displacement in the downstream direction of the dam crest induced by the Wenchuan earthquake were 19 and 25 mm, respectively. Several other earthquakes that have occurred since the end of construction have had insignificant effects on the dam. Dynamic analyses confirm that the dam performed and operated safely during and after the Wenchuan earthquake (Han et al., 2016).

Until December 2013, the Yele Dam had experienced eight cycles of water level rising and lowering. The left and right parts of the dam crest had moved in towards the centre of the valley with maximum displacements of 45 mm and 19 mm, respectively. The maximum vertical settlement was 184 mm at the downstream slope at el. 2,623.3 m.asl. at section D (see Figures 10.6 and 10.7).

In summary, the dam surface displacements show regular and consistent deformation patterns even after having experienced the Wenchuan earthquake, and the observed settlements are smaller than expected. This must mean that the rockfill and gravel are of high quality, the embankment was well compacted, and the alluvial overburden in the foundation is less compressible than anticipated.

10.4.6 Measured deformations inside the dam

In August 2007 (i.e., 19 months after the end of construction), observed settlements in the downstream fine transition zone are as shown in Figure 10.12. The maximum settlements, including the foundation settlements, at sections B, D, and F were 560, 1,060, and 550 mm, respectively. The maximum settlement at section D was nearly double that at sections B and F. Section D is

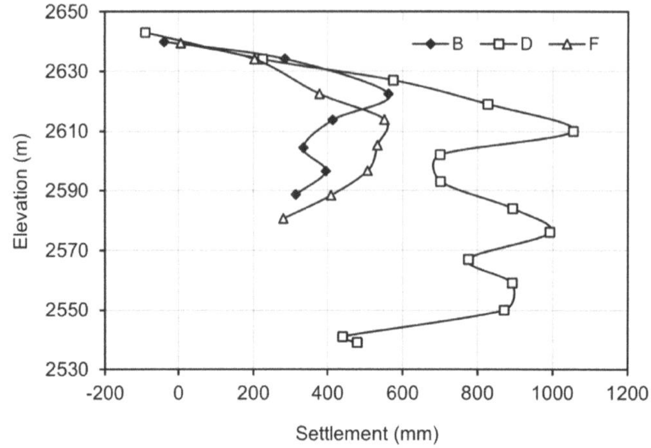

Figure 10.12 Settlements inside transition Zone 2 downstream of the core at three cross-sections. Observation date 3 August 2007.

the maximum cross-section and has a height of 124.5 m (Figures 10.6 and 10.7). Sections B and F (Figure 10.7) are located at the left and right banks, respectively, where the dam heights are about 60 and 70 m, respectively. The measured vertical strains in the transition zone adjacent to the core varied significantly, but the average compressive strains were 0.9%, 1.3%, and 0.8% in the three sections. In December 2013, the average compressive strains in the transition zone were 1.05%, 1.33%, and 0.8% in the three sections, only slightly different from the strain measurements in August 2007.

In August 2007, the maximum horizontal displacements normal to the dam axis in the downstream transition zone in cross-sections B, D, and F, were 30, 207, and 33 mm, respectively. The measurements of strains in the concrete plinth (as of January 2009) indicate very low values, and that no cracking has occurred. Strain metres are also installed in the cut-off wall at four sections. The observed results indicate low strains and satisfactory performance of the wall.

The measured settlement differences between the lower part of the asphalt core and the adjacent fine transition zones show large variations, but the asphalt core has settled more than the transition zones in all measurements. The maximum recorded settlement difference (November 2008) was about 60 mm. The settlement differences took place mainly during the dam construction period, and there is virtually no measured increase in differential settlements after the end of construction. Around 90% of the vertical strains in the core took place during the dam construction and impounding period, and there is minimal measured increase in the core strains during the subsequent 3-year operation period. The compressive vertical strains in the upstream and downstream sides of the asphalt core in the lower 25 m of the core wall in cross-section D were measured to be 2.5%–3%, while the average compressive strain in the downstream transition zone over the corresponding height was 1.7%. The May 2008 Wenchuan earthquake had insignificant effects on the strains in the core, and from November 2008 to December 2013, the settlements of the core and the settlement differences between the core and the adjacent transition zones did not change.

The vertical compression stresses measured by the total pressure cells installed at the bottom of the core are 0.65 MPa near cross-section A and 2.1 MPa near section D (see Figure 10.7). These values have stayed almost constant since the end of dam construction (December 2005) to the observation in November 2008, showing only very small variations caused by fluctuations in the reservoir level. The measured stresses are about 90% (section A) and 70% (section D) of the stresses computed by multiplying the local height of the core with the unit weight of the material above. Thus, the measured transfer of vertical stresses to the transition zones from the more compressible core is relatively small in the Yele Dam (i.e., little arching action).

Vertical strains in the core were back-analysed based on measured dam and foundation settlements in Figure 10.12 and the vertical compression stresses at the bottom of the core (Wang, 2008). The results from the back-analysis confirm that the most critical location for the asphalt core is 160 m from the left bank where the bottom of the cut-off wall leaves the rock base and goes over to the more compressible overburden (see Figure 10.9). The maximum shear strain in the most critical location is about 2.3%. Triaxial compression test results on specimens drilled out of the Yele Dam core show that the shear strain when significant cracking starts and failure begins is about 16.0% under a confining stress of 1.0 MPa. One may conclude that the computed strains (back-analysed) inside the Yele core are safely within a range where there is no concern about cracking.

The temperature in the core is observed in sections B, D, and F. During the first month after asphalt concrete placement, the temperature dropped from around 160°C to less than 20°C and gradually reduced with time. The temperature at the bottom of section D was 13.7°C in December 2004, 12.4°C in December 2005, 9.5°C in December 2007, and 9.4°C in November 2008. In July 2009, the temperatures at different points in the core were in the range of 7.2°C–12.5°C

Key references

Wang, W. 2008. Research on the suitability of asphalt concrete as water barrier in dams and dikes. PhD. Thesis, University of Oslo, Department of Geosciences, Norway.

Wang, W., Høeg, K., and Zhang, Y. 2010a. Design and performance of the Yele asphalt-core rockfill dam, *Canadian Geotechnical Journal*, 47(12):1365–1381.

10.5 Romaine-2 Main Dam, Canada

The Romaine-2 Main Dam with an installed nominal hydropower capacity of 640 MW is the biggest of the 4 power stations (Romaine-1 to Romaine-4) of the La Romaine Hydroelectric Complex in the province of Québec that provides a total of 1,500 MW. The reservoir is formed by the Romaine-2 Main Dam and five dikes (saddle dams) to create a reservoir area of approximately 81 km^2 at el. 243.8 m.asl. and has a storage capacity at the normal water level (NWL) of 3720×10^6m^3 (Alicescu et al., 2015; Smith, 2015).

10.5.1 Dam site conditions

The rock abutments mainly consist of sound granitoids varying from monzonite to quartz monzonite with average to coarse grains. The rock is generally sound. The banks of the river are characterized by the presence of outcropping, jointed rock, especially on the right side. In the area of the upstream cofferdam and the upstream shoulder of the dam, sub-horizontal open joints up to 500 mm wide were observed. No geological faults exist at the dam site.

For such geological conditions, the Romaine-2 Main Dam and the five dikes were selected to be rockfill embankments. Asphalt core dams would be more economical than earth core dams due to the long distance to reach suitable till borrow areas in the Romaine-2 region. Figure 10.13 shows the max cross-section for the Romaine-2 Dam. The cross-section is very similar to the one designed and constructed for the Storglomvatn Dam in Norway a few years earlier

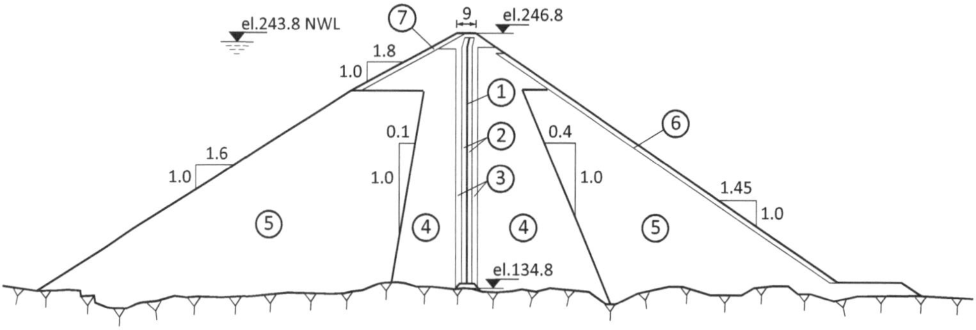

(1) Asphalt core; (2) fine transition zone, crushed rock, max. 80 mm; (3) coarse transition zone, screened rockfill or crushed rock, max. 200 mm; (4) quarried rockfill, max. 600 mm; (5) quarried rockfill, max. 1,200 mm; (6) protective layer; and (7) riprap, max. 1.000 mm, max. 25% <300 mm.

Figure 10.13 Cross-section of the Romaine-2 Main Dam at the highest (ICOLD, 2018).

(see Section 10.3). The Romaine-2 Main Dam is 112 m high with a crest length of 500 m, dam volume $4.1 \times 10^6 m^3$, and volume of asphalt core 18,850 m^3. The entire dam is founded on bedrock.

Six dams in the La Romaine Hydroelectric Complex were selected to be built as ACRDs. One of them (Dike F) was 84 m high and had a very steep right rock abutment of more than 70° (see Section 8.7). All the dams were completed on time and have performed very well.

10.5.2 Dam design and construction

The concrete plinth for the asphalt core was designed considering the bedrock conditions and the hydraulic head. The minimum width of the plinth is 3.9 m. When the hydraulic head (H) exceeded 78 m, the plinth width was determined by the ratio H/20 (see Section 8.4). The nominal thickness of the plinth was established at 0.45 m. However, due to the irregular bedrock profile along the dam axis, the thickness varied greatly, but it is nowhere less than 0.30 m. The asphalt core width is 0.85 m at the bottom and 0.50 m for the top 50 m. The core width varies step by step from the bottom to the top with the following widths: 0.85–0.75–0.65 and 0.50 m. Towards the abutments, the core is flared to double core width at that elevation. The core is vertical and centred on the dam axis, except at the top where there is an inclination of 1V:0.2H towards downstream mainly to facilitate the construction of the top part of the dam. The transition Zone 2 adjacent to the asphalt core is 1.5 m wide over the full height of the core.

The asphalt core and the adjacent fine transition zones (Zone 2) were placed in layers with a compacted thickness of 22.5 cm using 1.6-tonne and 3.9-tonne vibratory rollers, respectively. The coarse transition zones (Zone 3) were placed with a compacted layer thickness of 0.45 m using 15-tonne vibratory rollers with 4 passes. The rockfill in Zones 4 and 5 was placed with a layer thickness of 0.90 and 1.20 m, respectively, compacted by 15-tonne vibratory rollers with 8 passes, which resulted in very small shoulder deformations.

The bitumen grade used for the asphalt core is PG 52-34 with a penetration of 160/220 corresponding to grade B180 referred to in Chapter 3. The bitumen content is 6.9%. The aggregates in the asphalt concrete mix are limestone with a maximum particle size of 20 mm, and the filler content is 12%.

Construction works were scheduled from July 2011 to December 2013. Figure 10.14 shows the dam during construction. Due to the higher-than-expected construction rate, the dam was

Figure 10.14 Romaine-2 Main Dam during construction. (Courtesy of Vlad Alicescu, Hydro-Québec.)

completed 1 month early, saving Hydro-Québec considerable cost and enabling earlier energy generation than expected (Smith, 2015). The reservoir started being impounded in 2014. During the first month of impounding the reservoir level rose rapidly, about 70 m, and later much slower as the reservoir cross-section increased significantly. In November 2014, the reservoir water level reached el. 234.8 m.asl., which is normal water level (NWL). Figure 10.15 shows the 84-m high saddle dam (Dike F) in the Romaine-2 project. The right bank abutment was very steep and fractured, and the photo shows that the abutment had to be treated and cleaned to reduce the rockfall hazard and to prepare the foundation for the plinth (white area in photo).

10.5.3 Instrumentation and monitoring

Romaine-2 Main Dam is extensively instrumented and monitored (Smith, 2015), and the dam performance has been evaluated using a non-linear numerical stress-deformation analysis (Longtin et al., 2013).

The rockfill compression modulus for Zones 3, 4, and 5, back-analysed based on the measured settlements during construction, were found to be as high as 170, 180, and 80 MPa, respectively. The high modulus values measured for the rockfill 4 are due to the relatively low layer thickness during placement and the number of passes (8) by the 15-tonne roller.

After 1 year of operation, seepage flow measured at the downstream measuring weir was less than 2.5 L/second. The monitored results and field observations show that the core is virtually impervious, and the dam performance is very satisfactory.

Figure 10.15 Saddle dam (Dike F) in the Romaine-2 project. (Courtesy of Jean-Pierre Tournier, Hydro-Québec.)

Key references

Alicescu, V., Tournier, J. P., Kara, R., and Rosculet, D. 2015. Construction of La Romaine Complex in Northern Quebec: six years of great accomplishments. Behaviour of asphalt core dams. In *Canadian Dam Association Annual Conference*, Mississauga, Ontario, October.

ICOLD. 2018. *Asphalt Concrete Cores for Embankment Dams*. Bulletin 179. International Commission on Large Dams (ICOLD), Paris.

Smith, M. 2015. Rockfill settlement measurements and modelling of the Romaine-2 Dam during construction. In *25th ICOLD Congress*, Q.98, R.31, Stavanger, Norway.

Longtin, H., Verret, D., James, M. Grenier, S., Noel, J-F., and Hammamji, Y. 2013. Romaine-2: numerical stress-deformation analyses of the dam and dykes. *CDA 2013 Annual Conference*, Quebec, Canada.

10.6 Zarema Dam, Ethiopia

The Zarema Dam is in northern Tigray, forming a huge water reservoir for sugar cane irrigation. The reservoir area is nearly 75,000 ha. made for the Wolkyte Sugar Development Project in Ethiopia (Sembenelli and Afsari-rad, 2015).

10.6.1 Dam site conditions

The valley for the dam site is about 200 m deep, and the cross-section of the valley has a nearly symmetrical "U" shape. The width of the valley is about 200 m at riverbed level and about 900 m at el. 946.4 m.asl. at normal water level (NWL). The bank slopes are close to 1V:2H on the right side and 1V:2.5H on the left side.

The rock along the dam axis consists predominantly of a metamorphic formation. The metamorphic rock is strong, characterized by numerous, closely spaced joints. Due to erosion, the rock has been decompressed, and local faulting has created bands of fracturing. The weathering of the metamorphic is intense at the highest elevations, and it gradually decreases deeper down towards the fresh rock. Closer to the riverbed, the rock joints are tighter, and the rock permeability decreases. The water table is around el. 815 m.asl., which corresponds to the river level.

The sedimentary formation is above the dam crest elevation at el. 950 m.asl. The top 10–15 m of the metamorphic rock has been intensely weathered prior to the deposition of the sedimentary and has the consistency of soil locally. Therefore, a 25-m thick sedimentary formation covers the weathered metamorphic and forms the plateau flanking the river. A mantle of alluvium and scree, up to 10 m thick, covers the banks, and about 15-m thick alluvial bed fills the bottom of the valley.

The Zarema Dam is in a region with low seismicity. The high Simien Mountain range ensures relatively high runoff in the Zarema River at the dam site, and the rainy season is from June to September. The annual average runoff is around $1.3 \times 10^9 m^3$, and the expected total inflow in July and August roughly corresponds to 80% of the mean annual inflow.

10.6.2 Dam core design and special construction procedure

After preliminary investigations had been completed, and various design options considered, the Zarema Dam was designed and built as an ACED. The need for an 80 m high cofferdam with a volume of about $2.0 \times 10^6 m^3$ made it important for both practical and economic reasons to incorporate the cofferdam in the main dam.

As originally designed, the cofferdam intersected a fault zone in the left abutment and filled a deep lateral valley, thus requiring a fill volume much larger than necessary. As fill volume and required construction time are of the essence when building a cofferdam, the cofferdam axis was moved about 35 m downstream towards a narrower and stronger section of the valley. The reduction in cofferdam fill volume obtained with such a shift was estimated to be close to 30%. This design change would have required a shift of the axis of the main dam by a similar distance that implied an increase of nearly $1.6 \times 10^6 m^3$ in the volume of rockfill placed downstream of the asphalt core. To efficiently incorporate the excavation and the grouting that had already been done along the asphalt core trench for the original design, the core axis was not changed. On the contrary, the downstream buttressing of the core against the hydrostatic thrust was improved by the larger embankment volume existing downstream of the core. The top 15 m of the originally designed vertical asphalt core was no longer required as the designer chose to use a polymer geomembrane facing on the upstream slope of the top part of the dam as an "extension" of the vertical asphalt core.

Figure 10.16 shows the re-designed dam cross-section at the maximum dam height. The dam height is 152 m with a 135 m high vertical asphalt core in the cross-section shown in Figure 10.15. The crest lengths of the dam and of the crest of the asphalt core are 805 and 695 m, respectively. The dam is designed with an upstream slope of 1V:1.75H up to the top of the cofferdam. A 28-m wide platform is designed on the top of the cofferdam providing a river crossing during construction. It is connected with the slope of 1V:1.6H for the upstream embankment up to el. 939 m.asl. At this elevation, there is a 12-m wide platform at the top of the core that is above the minimum operating water level at el. 937.75 m.asl. The platform facilitates future maintenance of the core or to the top facing.

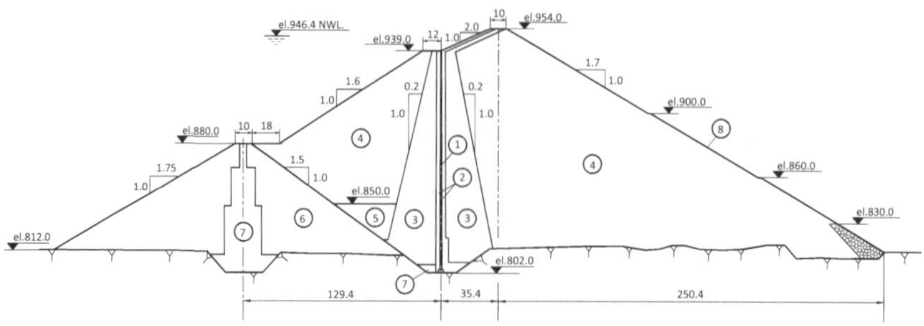

(1) Asphalt core, compacted in 0.2 m layers; ② fine transition zone, alluvial gravel (<60 mm), compacted in 0.2 m layers; ③ quarried sedimentary rock (<200 mm), compacted in 0.4 m layers; ④ quarried sedimentary rock (<600 mm), compacted in 0.8 m layers; ⑤ alluvium from river bed, compacted in 0.6 m layers; ⑥ upstream cofferdam; ⑦ red soil, compacted in 0.2 m layers; and ⑧ riprap.

Figure 10.16 Zarema Dam cross-section at maximum dam height (from Sembenelli and Afsari-rad, 2015).

The top 15-m upstream part of the dam is designed with a slope of 1V:2.0H and is sealed with a polymer geomembrane facing. The downstream slope of the dam is 1V:1.7H, and two 2-m wide berms are located at el. 860.0 m.asl. and el. 900.0 m.asl., respectively. The volume of the dam is about $14.5 \times 10^6 m^3$ with a volume of 53,900 m^3 for the asphalt core.

The deepest elevation of the core trench in the riverbed is el. 802.0 m.asl. Cuts on both sides of the core trench were set at variable slopes between 1V:0.9H and 1V:1.3H. Cuts in the riverbed alluvium were set at a slope 1V:1.5H to provide adequate stability and to minimize the need for support.

The asphalt core sits on a 6-m wide and 0.8 m thick concrete plinth cast on the foundation bedrock. The plinth profile is made without large angle changes between adjacent segments of the plinth (see Section 8.4). A steel wire mesh reinforces the surface of the plinth. Construction joints are introduced in the plinth at about 20 m distance. Blanket grouting consists of six rows under the plinth, and the deep curtain grouting consists of two rows down to 60 m below the plinth.

The lowest elevation for the asphalt core is el. 804.25 m.asl., and the first two asphalt concrete layers (about 0.4 m) serve as the core base 2.0 m wide. Above the core base, the core width is 1.2 m constant up to el. 860.0 m.asl. changed to 1.0 m up to el. 900.0 m.asl., and changed to 0.8 m up to the top of core at el. 939.0 m.asl. The fine transition zones adjacent to the core are 1.5 m wide. A 2-m thick plug of red soil is placed and compacted upstream of the core transition zone. A 4.5-m wide transition and drain zone is provided up to el. 930.0 m.asl. over the whole bottom of the core trench.

The construction of the Zarema Dam started in late 2013, and the 80-m high cofferdam was completed before the onset of the rainy season in 2014. The plinth and the grout curtain in the riverbed were completed in February 2015. The testing of materials, mix design, and field test section were carried out from January to March 2015. The bitumen content (B70) was 6.4%. The mix for the asphalt concrete in the core is shown in Table 10.4. A 10–20 mm thick sandy asphalt mastic was placed on the plinth surface.

The asphalt core and fine transition zones were constructed layer by layer with a compacted layer thickness of about 0.2 m. The 2.0-m wide asphalt core base was compacted by two

Table 10.4 Asphalt concrete mix used for the Zarema Dam core

	Sieve size (mm)				*Filler (<0.075 mm)*		*Bitumen (percentage of total weight)*
	12–19	*6–12*	*3–6*	*0–3*	*Limestone*	*Cement*	
Ratio (%) of mineral weight	17	23	21	26	6.5	6.5	6.4

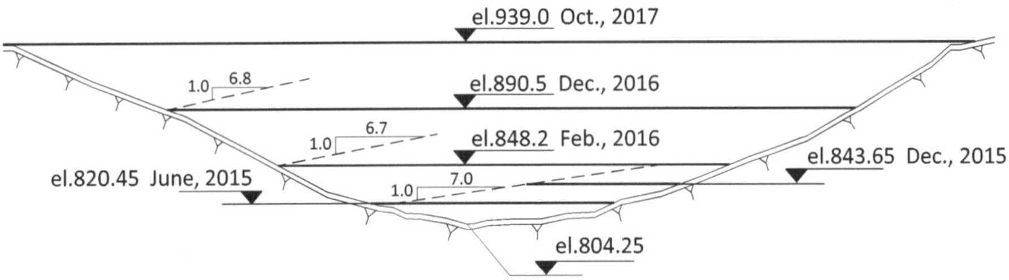

Figure 10.17 Longitudinal section of dam at various construction stages of core.

1.6-tonne vibratory rollers with a drum width of 0.9 m. When the core width was 1.2 or 1.0 m wide, the asphalt core was compacted by a 3.5-tonne vibratory roller with a drum width of 1.3 m. When the core width was 0.8 m, the asphalt core was compacted by a 1.6-tonne vibratory roller with a drum width of 0.9 m. The fine transition zones (Zone 2) were compacted by 3.5-tonne vibratory rollers with a drum width of 1.3 m. Figure 10.17 shows the construction progress of the asphalt core at different stages in the longitudinal cross-section along the core axis.

The dashed lines in Figure 10.17 indicate that the asphalt core was first constructed towards the right abutment. The area along the left abutment temporarily stayed behind to be able to complete the foundation grouting in the left part of the foundation and abutment. The temporary construction joint between the right and left sides of the core had a gentle slope of about 1V:7H to ensure stability and a tight construction joint between the left and right side after placing and roller compaction of the layers on the left side.

10.6.3 Core construction method and scheduling

The construction of the asphalt core started in early April 2015 at el. 804.25 m.asl. and reached el. 820.45 m.asl. in early June 2015. Figure 10.18 shows the overview of the dam under construction in May 2015. The asphalt core was not constructed from July to September 2015 during the rainy season. Another reason for the stoppage of asphalt core construction was that the foundation at the left abutment needed to be completed by additional curtain grouting before placing and compacting asphalt layers.

The construction of the asphalt core restarted in October 2015. As the additional work of the foundation curtain grouting in the lower part of the left abutment had not been completed yet, the asphalt core had to be constructed towards right abutment leaving a slope of 1V:7H along the core longitudinal direction. In December 2015, the asphalt core reached el. 843.65 m. asl. in the right section. Figure 10.19 shows the asphalt core under construction in the right

Figure 10.18 Overview of the Zarema Dam under construction in May 2015, looking towards left abutment.

section along the core longitudinal direction in November 2015. After the additional work of the foundation grouting in the lower part along the left abutment had been completed, the construction of the asphalt core in the left section restarted from el. 820.45 m.asl. In February 2016, the asphalt core in the left and right sections reached the same elevation at el. 848.2 m. asl. The foundation in the left abutment needed grouting in the middle and upper parts in two stages along the left abutment. The asphalt core had to be constructed towards the right abutment again, and two slopes of about 1V:7H along the core longitudinal direction were left in the middle and upper parts in the left section, respectively. In October 2017, the core reached the design top at el. 939.0 m.asl. Figure 10.20 shows the asphalt core near the end of dam construction in October 2017.

Key references and acknowledgements

Sembenelli, P., and Afsari-rad, V. 2015. High asphalt core for Zarema Dam. *Proceedings of the 25th ICOLD Congress*, R98-Q34, Stavanger, Norway; pp 526–538.

The information presented in this section is taken from the published article by Sembenelli and Afsari-rad (2015) and from the project reports by PowerChina Jiangxi Hydropower Engineering Bureau Co., Ltd. who constructed the asphalt core for the Zarema Dam with the technical assistances from the Xi'an Huize Construction Engineering Co. Ltd., China. Their support and permission to publish the data are gratefully acknowledged.

Figure 10.19 Asphalt core under construction for the Zarema Dam in the right section. Photo shows the slope of 1V:7H along the core axis in the longitudinal direction (November 2015), looking upstream and at right abutment.

Figure 10.20 Asphalt core for the Zarema Dam near the end of construction in October 2017.

10.7 Quxue Dam, China

The Quxue Dam is on the Shuoqu River, which is the largest tributary of the Dingqu River. The dam is in Derong County, Sichuan Province, and the dam site is about 126 km from Shangri-la County Town.

10.7.1 Dam site conditions

The dam site is in a mountainous terrain, and the cross-section of the narrow valley (canyon) appears as an asymmetrical "V" shape (Figure 10.21). The average slope of the left bank is about 70°, and the slope of the right bank is also very steep, especially towards the top. The width of the valley is about 200 m at the normal water storage level (NWL el. 2,330 m.asl.), and the ratio of valley width to valley depth at that level is about 1.3.

There are no primary regional faults crossing the dam foundation, but there is one main secondary fault joint in the left abutment (shown as no.7 in Figure 10.24) with a fractured zone width of 0.3–1.0 m. In the right bank, there are two groups of various size karst caves. The first group (shown as no.8 in Figure 10.24) intersects the river at a small angle. The caves are interconnected and filled with greyish-white calcareous and purple-red mud. The second group of three karst caves are exposed on the surface and are located downstream about 260 m from the dam axis. The main potential leakage paths at the dam site are solution cracks and channels and some leakages from structural planes of weakness in the rock.

Figure 10.21 Quxue Dam site, looking upstream.

The depth of alluvial river overburden above bedrock is about 30 m at the dam site. The exposed rock that is marbleized, fine-grained limestone, and the saturated UCS of the moderately weathered rock was found to be between 40 and 60 MPa. The depth of the weathered rock is 11–35 m in the riverbed, 12–63 m in the left bank, and 7–65 m in the right bank. Fissures and cracks have developed due to unloading in the deep cut with high and steep banks.

10.7.2 *Dam type selection and design*

For the topographical shape and the geological conditions at the site, a roller-compacted concrete (RCC) arch dam, a CFRD, and an ACRD were examined and compared. To decide among the three options, an emphasis was placed on required abutment stabilization and foundation treatment.

The selected dam site for the RCC arch dam was about 350 m upstream from the site for the CFRD and ACRD. At the arch dam location, the cross-section of the dam site has an almost symmetrical "V" shape that simplified the overall layout arrangement for the RCC arch. However, there are karst caves and intensely weathered and unloaded rock in the right bank that would require extensive excavation works. The excavation works and the long transport of materials made the RCC arch dam the most expensive among the three options (Wang et al., 2017b).

The steep left bank would make it difficult for the arrangement and construction of the perimeter plinth for the face slabs of a CFRD. Furthermore, potential cracking might develop in the sloping face slabs due to the anticipated uneven and differential settlements of the embankment.

The construction on the steep left abutment is simpler for the plinth under a vertical asphalt core for the ACRD. Asphalt core flexibility and ductility allow the core to adjust to embankment deformations without detrimental cracking. In addition to the simpler treatment of the steep left abutment, the ACRD had the advantage of a shorter construction period, lower construction costs, and earlier electricity generation. The dam type evaluation process and comparisons ended up with selecting the ACRD option.

Figure 10.22 shows the ACRD cross-section that was selected. The asphalt core shown in that cross-section is 132 m high from el. 2,201 m.asl. to el. 2,333 m.asl., and the total dam height is

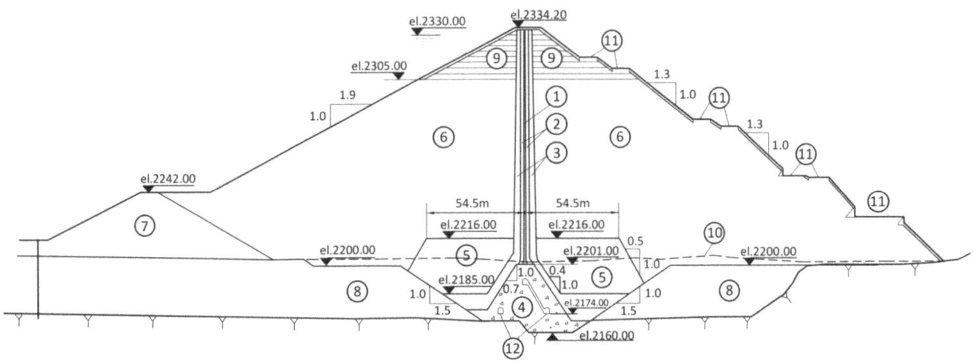

① Asphalt core; ② fine transition zone; ③ coarse transition zone; ④ concrete monolith; ⑤ well-compacted rockfill; ⑥ rockfill; ⑦ upstream cofferdam; ⑧ alluvial overburden; ⑨ bi-directional polypropylene geo-grids; ⑩ natural ground surface; (11) 10 m wide road berm; and (12) transportation and grouting galleries.

Figure 10.22 Cross-section A of the Quxue Dam (location of cross-section A is shown in Figure 10.26).

174 m from el. 2,160 m.asl. to el. 2,334 m.asl. The crest length for the dam is 220 m. Mainly due to the seismicity of the region, the dam is designed with a gentle upstream slope of 1V:1.9H, an average downstream slope of 1V:1.84H, and a wide crest of 15 m. The downstream slopes between the 10 m wide road berms are steep, 1V:1.3H. As an earthquake-resistant measure, bi-directional polypropylene geo-grids with a vertical distance of 2.4 m between the grids were placed horizontally to reinforce the top 30 m of the dam (from el. 2,305 m.asl. to the dam crest at el. 2,334 m.asl.) (see Chapter 9).

The vertical asphalt core, located 2.0 m upstream of the dam centreline, is 1.5 m wide at the bottom and decreasing gradually to 0.6 m at the top (el. 2,333 m.asl.). Comprehensive investigations were conducted on the available materials and on various mixes for the asphalt core. Bitumen grade B70 was selected. The aggregates were crushed limestone, and the added filler consisted of limestone powder. The total filler content is 13% and the bitumen content 6.4%.

10.7.3 Core and plinth design

The bottom of the core wall is flared out against the foundation concrete plinth on top of the concrete monolith to a width of 3.0 m at the core-plinth interface. Similarly, the core is flared out against the plinth at the interface with the steep abutments to twice the core width. Figure 10.23 shows the surface of the foundation plinth and the plinth along the left abutment. To further ensure that the connection of the asphalt core to the plinth is watertight, a row of copper waterstops was installed along the core axis. The surface of the plinth under the core was made concave with a maximum depth of 250 mm at the core centreline. The top width of the foundation plinth is 8 m, and the width of the plinth anchored in the rock along the abutments is 8 m at the bottom and is gradually reduced to 5 m at the top. To provide a smooth abutment profile for the asphalt core and a stable concrete plinth along the steep rock abutments, the plinth was designed to be embedded in the rock abutments (Figure 10.23). For that purpose, the thickness of the plinth normal to the slope varies from 3 to 5 m to obtain a satisfactory plinth surface profile along the rough and uneven abutments.

The width of the upstream and downstream fine transition zones (Zone 2) adjacent to the asphalt core is 2.0 m, and the width of the upstream and downstream coarse transition zones (Zone 3) is 4.0 m at the bottom and 2.0 m at the top. To reduce the settlements and deformations of the rockfill in the lower central part of the dam, especially well-compacted rockfill (Zone 5 in Figure 10.22) was placed adjacent to the upstream and downstream coarse transition zones.

Figure 10.24 shows the longitudinal section along the asphalt core axis and the layout of the grouting galleries and grout curtains in the foundation.

10.7.4 Dam construction, quality control, and reservoir impounding

The construction of the concrete monolith across the river channel started 28 January 2015 and was completed 31 May 2015. Embankment fill construction started 13 April 2015, and the top of the asphalt core was reached 27 February 2017 at el. 2333.35 m.asl. The design elevation for the top of the core is el. 2333.0 m.asl., but the core was built to el. 2333.35 m.asl. to compensate for the estimated settlement after construction. Zoning, compaction specifications, and quality control for the dam are described in Table 10.5.

The reservoir impounding started on 12 February 2017 from el. 2,208 m.asl., and on 17 June 2017, the water level reached el. 2,320 m.asl., which is 10 m below the NWL at el. 2,330 m.asl. The water level at el. 2,320 m.asl. was kept during the flood season from June to October. On 30

Figure 10.23 View of the surface of the foundation plinth (with black bitumen coating) and the plinth started up along the left abutment.

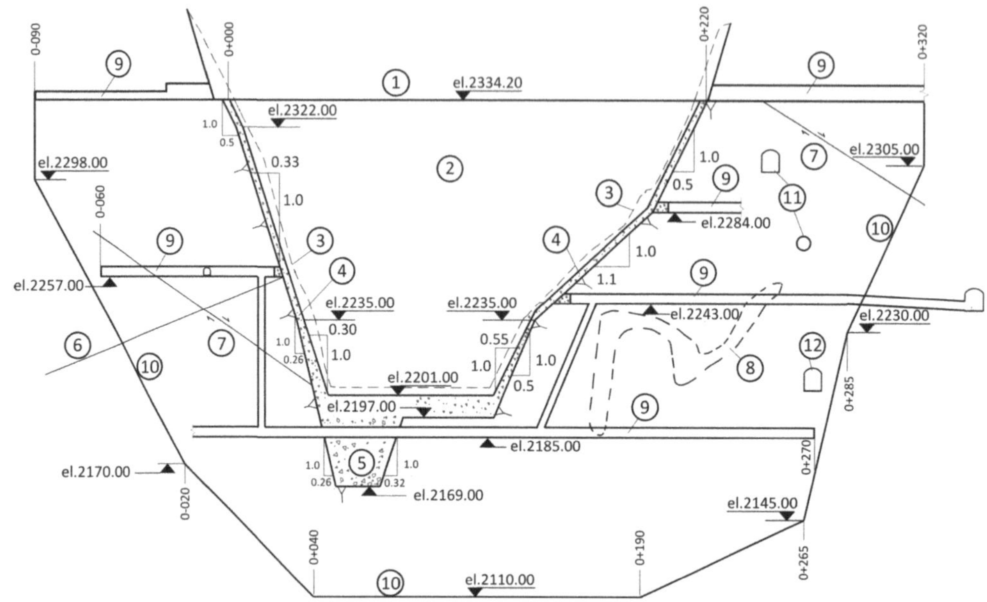

(1) Dam crest; (2) asphalt core; (3) ground surface; (4) concrete plinth; (5) base concrete monolith; (6) rock joint; (7) fault; (8) karst caves filled with concrete after being washed; (9) grouting and transportation gallery; (10) bottom level of grout curtain; (11) spillway tunnel; and (12) diversion tunnel.

Figure 10.24 Longitudinal section along the asphalt core axis showing grouting galleries in the foundation.

July 2017, the Quxue hydropower station started to generate electricity. In November 2017, the water level reached el. 2,327 m.asl. only 3 m below the NWL. Figure 10.25 shows the progress of dam construction, impounding, and operation until November 2017.

10.7.5 Dam monitoring and performance observations

An extensive field instrumentation and monitoring program is implemented for the Quxue Dam (Feng et al., 2020a). Figure 10.26 shows an overview of the dam with the locations of cross-sections A, B, and C. The distances from the left bank are 70, 140, and 40 m for the cross-sections A, B, and C, respectively. All the measurements inside the dam are collected at the five observation huts on the downstream dam slope and through pipes up to the dam crest.

The vertical settlements inside the dam are obtained by water level gauges, and the horizontal displacements are obtained by extensometer measurements relative to the movements recorded at the five observation huts at the downstream slope of the dam (no.11 in Figure 10.26). There is a complete record of lateral core displacements since the start of reservoir impounding.

10.7.6 Horizontal displacements of asphalt core since start of impounding

Figure 10.27 shows the locations of inclinometers, settlement difference metres, strain metres, thermometers, and piezometers over the dam height in cross-section A. There are similar arrangements in cross-sections B and C.

Table 10.5 Zoning, compaction specifications, and quality control for the Quxue Dam

Zone (see Figure 10.21)	*Material*	*Layer thickness (m)*	*Compaction by vibratory roller*	*Quality control*
Thickness asphalt core, 0.6 m (top)-1.5 m (bottom)	Bitumen B70 Aggregate (0–19 mm)	0.25	8 passes, 3 tonnes	Air void content < 3.0%
Fine transition Zone 2, width 2 m	Crushed rock (0–60 mm), grain size < 5 mm passing 10%; grain size < 0.075 mm passing < 5%	0.25	8 passes, 3 tonnes	Dry density ≥ 23.0 kN/m³, porosity ≤ 20%,
Coarse transition zone 3, width 2 m (top) - 4m (bottom)	Crushed rock (0–150 mm), grain size < 5 mm passing15%–25%; grain size < 0.075 mm passing <5%	0.5	8 passes, 26 tonnes	Dry density ≥ 22.8kN/m³, porosity ≤ 21%
Rockfill	Quarried rock (0–800 mm), grain size < 10 mm passing10%–25%; grain size < 5 mm,	1.0–1.2	10 passes, 33 tonnes	Dry density ≥ 23.0 kN/m³, porosity ≤ 21%
Well-compacted rockfill	passing <15%; grain size < 0.075 mm passing <5%	1.0	12 passes, 33 tonnes	Dry density ≥ 23.3 kN/m³, porosity ≤ 19%

Shape-Accel-Array (SAA) type inclinometers, installed on the downstream face of the asphalt core in cross-sections A, B and C, measure the displacements of the core in the upstream-downstream and longitudinal (left bank-right bank) directions since the start of reservoir impounding. The casings for the inclinometers were installed on the downstream face of the core during dam construction, and the SAA sensors were put into the casings from the dam top after dam construction. Unfortunately, the casings in cross-sections B and C were clogged at el. 2,269.0 m. asl. and el. 2,257.0 m.asl., respectively, by accidentally falling stones, so the SAA sensors were unable to reach the bottom elevations through the casings. Figure 10.28 shows the measured displacements.

For cross-section A in Figure 10.28, the results show that when the reservoir water level reached el. 2,320 m.asl. in June 2017, the top of the core had moved in the upstream direction about 20 mm. In November 2017 with water level at el. 2,327 m.asl., the top of the core had moved in the downstream direction by about 15 mm caused by downstream creep and the water level increase of 7 m. Most of the core moved downstream with a maximum of about 90 mm at el. 2,230 m.asl. It should be noted that the displacements of the core presented in cross-sections B and C are the relative displacements to the elevations where the casings were clogged, which are not the levels of casing fixity. For cross-section B, the core moved in the

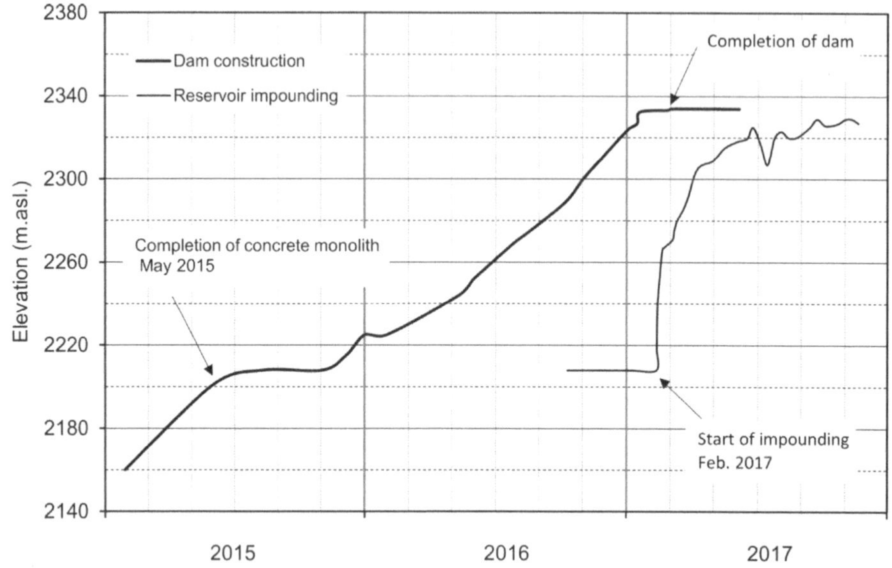

Figure 10.25 Progress of dam construction and reservoir impounding.

(1) Inlet of hydropower tunnel; (2) inlet of spillway tunnel;(3) water level at el. 2,327 m.asl., which is 3 m below the NWL at el. 2,330 m.asl.; (4) exit and entrance of transportation tunnel at left dam crest; (5) 15-m wide dam crest at el. 2334.2 m.asl.; (6) exit and entrance of transportation tunnel at right dam crest; (7) 10-m wide road berms; (8) downstream slope 1V:1.3H between road berms (average slope 1V:1.84H); (9) platforms for vehicle "U" turn; (10) exit and entrance of transportation tunnel and platform for vehicles "U" turn; and (11) observation huts.

Figure 10.26 Overview of the Quxue Dam and locations of cross-sections A, B, and C.

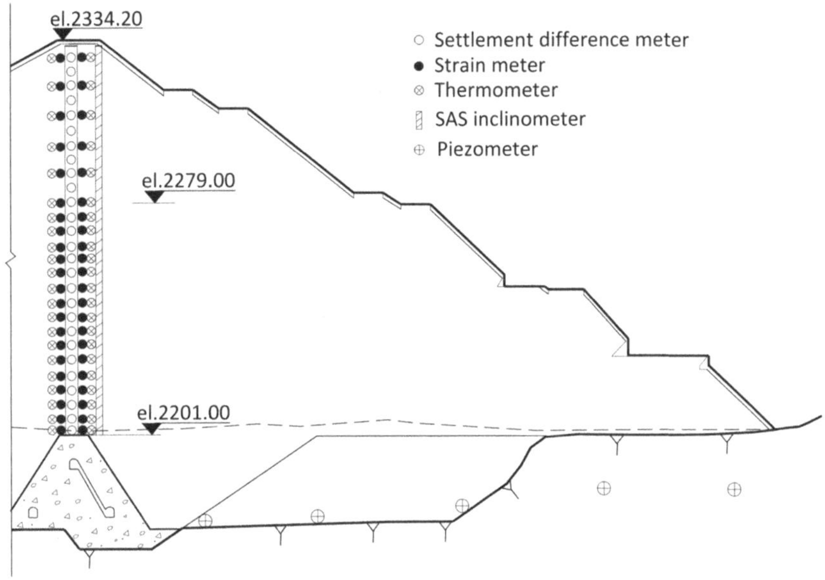

Figure 10.27 Locations of inclinometers, settlement difference metres, strain metres, thermometers, and piezometers in cross-section A.

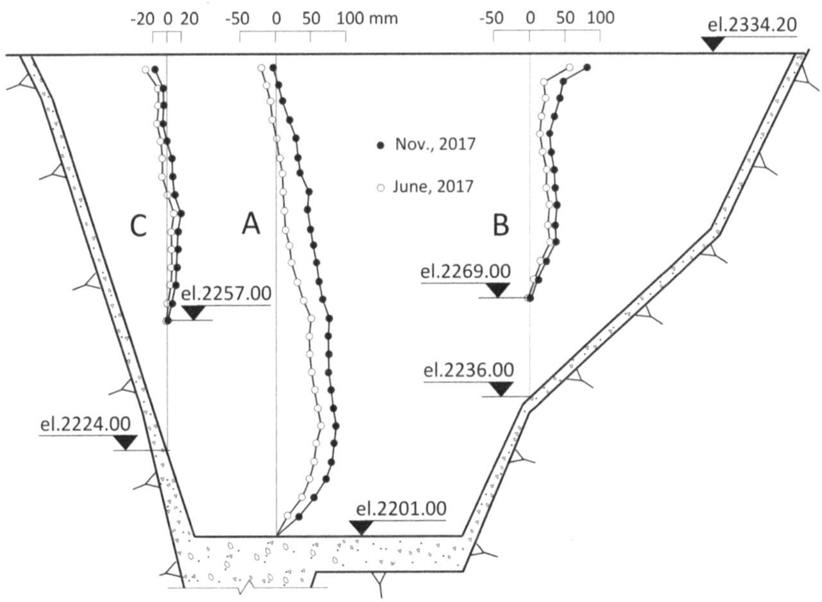

Figure 10.28 Measured displacements of the asphalt core at three cross-sections since start of reservoir impounding. Displacements in the upstream-downstream direction (positive towards downstream).

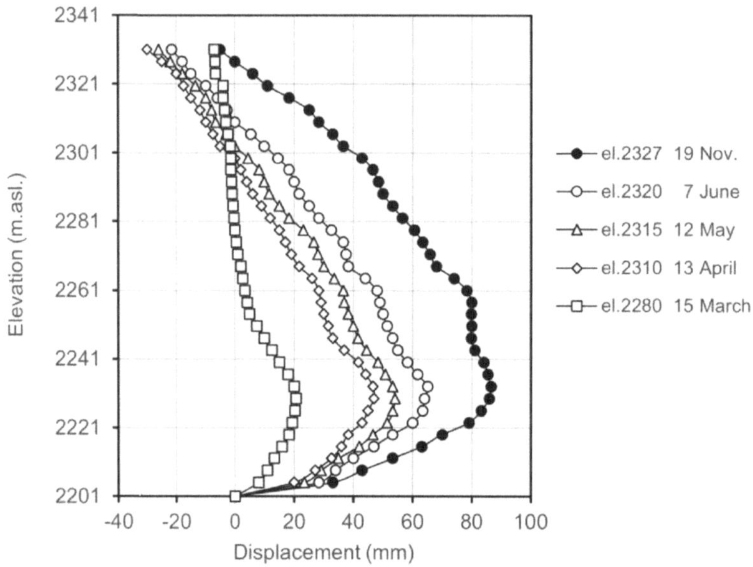

Figure 10.29 Measured horizontal core displacements in the upstream-downstream direction during reservoir impounding at different reservoir water levels in cross-section A (positive towards downstream).

downstream direction with a maximum of about 80 mm at the top of the core in November 2017. The top of the core in cross-section C first moved upstream about 25 mm (June 2017) and then moved in the downstream direction a small amount and was 20 mm in the upstream direction by November 2017. There were very small displacements measured of the core in cross-section A in the left-right direction since start of reservoir impounding.

Figure 10.29 shows in more detail the development of horizontal core displacements since the start of reservoir impounding in cross-section A.

As expected, the upstream shoulder experienced settlements due to the wetting effect during impounding and that caused the top part of the core to first move in the upstream direction before it later moved in the downstream direction.

The reservoir impounding began on 12 February 2017 from el. 2,208 m.asl. and reached el. 2,280 m.asl. on 15 March. The water level rose as much as 72 m in the 31 days. This caused the core to move in the downstream direction about 20 mm at el. 2,230 m.asl. However, above the water level at el. 2,280 m.asl., the core had moved in the upstream direction by about 8 mm at the top of the core. The asphalt core seats on top of the concrete monolith, and there is about 35-m deep compacted rockfill and alluvial overburden behind the monolith (Figure 10.22, Zones 5 and 8). Settlements in the deep rockfill and alluvial overburden contributed to the large lateral displacements towards the bottom of the core (Figure 10.28). When the water level reached el. 2,310 m.asl. on 13 April, the core had moved downstream with a maximum displacement of about 45 mm at el. 2,230 m.asl. and upstream with a maximum displacement of about 30 mm at the top of the core. When the water level reached el. 2,327 m.asl. on 19 November, the lateral displacement curve shows a maximum downstream displacement of about 90 mm at el. 2,230 m.asl. and an upstream displacement of 5 mm at the top of the core.

10.7.7 Settlement differences between the core and transition zone 2

Special gauges installed at the upstream and downstream interfaces between the core and the fine transition zones measured the settlement differences between the core and the fine transition zones at different elevations in cross-sections A, B, and C. The special gauges are vibrating wire settlement difference metres with a maximum range of 150 mm. The sensor of the metre between the two ends measures the vertical settlement difference between the core and the fine transition zone. The clear distance of the metre between the core face and the measurement point in the transition zone was about 150 mm. The measurement points were arranged with an individual height difference of 5 m below el. 2,279 m.asl. and 10 m above that elevation.

The measurements show that there are significant variations in the locally measured settlement differences over the core height. When the gauges were installed during construction, the asphalt core was still hot and soft. The asphalt core is in contact with Zone 2 material with a particle size of 0–60 mm, and the particle size may have caused significant variations in the locally measured settlement differences. In general, the asphalt core settled more than the transition zone, and the settlement differences took place mostly during the dam construction period. There is a relatively insignificant increase in the settlement differences due to reservoir impounding and creep. The measured settlement differences are in general agreement with the findings of the Maopingxi ACED and the Yele ACED (Wang et al., 2010a).

Most of the measured local settlement differences are less than 20 mm, and the magnitudes of the settlement differences do not seem to be related to the height of the measuring point above the base of the core. Only a few of the settlement differences were larger than 20 mm, but all less than 30 mm.

The asphalt core and the adjacent fine transition zones on both sides of the core are placed in the same construction operation with a compacted layer thickness of 20–25 cm. The interface on the upstream and downstream sides of the core will in practice show a zigzag pattern due to the method of construction (see Figure 6.15). The asphalt core-gravel interface is strongly interlocked, but with time, the core tends to settle more than the transition zones due to the core creep behaviour. Field measurements from other ACEDs show maximum settlement differences between the asphalt core and transition zones to be in the range of 20–50 mm (Wang et al., 2010a). The differential settlements are distributed over the distance of about 150 mm between the core face and the measurement point in the fine transition zone. Laboratory model test results confirm that the asphalt core face does not detach from the fine gravel transition zone even for large shear displacements (Wang et al., 2018). Since the core has a lower compression modulus and tends to "hang" on the adjacent fine transition zones, the vertical stresses in the core are lower than the vertical overburden stress. This arching effect has been measured in several ACEDs (Wang et al., 2010a) and is discussed in Chapter 8.

10.7.8 Temperature reduction with time in the asphalt core

Forty-four (44) thermometers were installed in the asphalt core in cross-sections A and B (Figures 10.26 and 10.27) with an individual height difference of 5 m to cover the whole height of the core. Twenty-four (24) thermometers were installed in three consecutive layers placed during construction and in four different locations in the lower part of the asphalt core.

Table 10.6 Temperature in asphalt core with time after placement of asphalt layer (°C)

Time (month)	*1/3*	*2/3*	*1*	*2*	*3*	*4*	*6*	*7*	*8*	*9*
Range	40–68	30–55	27–48	24–38	18–36	16–35	15–33	15–32	12–30	12–25
Average	53.5	43.7	36.8	31.7	26.5	23.4	21.4	19.6	18.4	17.5

The thermometers started to record the temperature in the core immediately after they were installed at different elevations. The measured results are shown in Table 10.6.

The temperature of the asphalt core was in the range of 145°C–165°C during placing and compaction and dropped to a range of 60°C–100°C in a few days. The temperature reduction in the asphalt during the first days after placement is strongly dependent on the weather conditions and how long the core surface is exposed before the next layer is placed. Once the asphalt is covered by the next asphalt layer, the temperature in the previous layer reduces at a very slow rate.

For the Quxue Dam asphalt core, Table 10.6 shows that the average temperature was about 37°C after 1 month and about 21°C half a year after the asphalt layer had been placed. The temperatures measured in each of three consecutive asphalt layers followed almost the same temperature reduction curve. During reservoir impounding, the temperature in the asphalt core dropped to a range of 13°C–18°C. In November 2017, 9 months after end of dam construction and just after first impoundment, the average temperature over the height of the asphalt core was 15.3°C.

10.7.9 Performance of concrete plinth under the core

The concrete plinth along the abutments was placed on a clean and fresh surface of rock and was built in slabs with slab lengths of 7–10 m along the core centreline (Figures10.22 and 10.23). The construction joints between the slabs were sealed with rubber and copper water stops.

The displacements of the plinth, the joints between the slabs, and the shear displacements of the core along the plinth were monitored. The one-way joint metres are vibrating wire displacement metres with a maximum survey range of up to 25 mm. The metre measures the joint opening or closing between two adjacent slabs. To measure the shear displacement between the concrete plinth surface and the base of the asphalt core, one end of the metre is installed in the plinth surface while the other end is placed in the asphalt core.

The measurements indicate that the shear displacements at the interface between the core base and plinth along the left and right abutments were less than 10 mm. Most of the shear displacements took place during dam construction. Along the interface, the middle and lower part of the core sheared downwards, while it sheared upwards towards the top of the core. The small shear displacements have no effect on the watertightness of the connection between the core and the plinth. This is also documented by laboratory tests on the connection between core and plinth when using sandy asphalt mastic at the interface (Wang et al., 2010a, 2017a).

The measurements indicate that the joints between adjacent slabs along the plinth closed or opened less than 3 mm. The small joint openings have no effects on the joint water tightness as the joints are provided with rubber and copper water stops.

Key references

Feng, S., Wang, W., Hu, W., Deng, Y., Yang, J., Wu, S., Zhang, C., and Høeg, K. 2020a. Design and performance of the Quxue asphalt-core rockfill dam, *Soils and Foundations*, 60(2020): 1036–1049.

Wang, W., Zhang, Y., and Xu, T. 2017b. Design and construction of the Quxue asphalt core rockfill dam, *Hydropower and Dams*, 24(4): 87–92.

10.8 Numerical analysis of the Quxue asphalt core dam

The design, construction, and field monitoring of the 174-m high Quxue ACED are presented and discussed in Section 10.7. Comparisons between the field measurements and results obtained by non-linear numerical stress-strain analyses are presented in this section that also compares the computed results obtained by 2D and 3D analyses for this dam in a narrow canyon.

Triaxial compression creep tests were conducted on asphalt specimens (diameter of 100 mm and 200 mm in height) drilled out of the Quxue Dam core to obtain deformation and strength parameters for the asphalt core material model (Wang and Høeg, 2016). The rockfill and gravel material parameters for the dam embankment were determined by triaxial compression tests on scaled-down rock material and by back-calculation of rockfill compression modulus to match the vertical settlements measured during construction.

Based on an evaluation of the field measurements, the laboratory test results, and by performing static numerical analyses, the authors have summarized the following findings and conclusions (Wang et al, 2017b; Feng et al., 2020a; Qiu et al., 2021):

- The 3D finite element analysis proved very useful when interpreting the dam and core performance. The comparison with the results from the 2D analysis showed, as expected, that the effects on the dam and core behaviour of the narrow valley and steep abutments are very significant and essential to consider in the design analysis and performance evaluation.
- At the end of construction (EoC), the maximum settlement occurring at about mid-height of the dam is only approximately 0.7 m in the core and 0.6 m in the upstream and downstream dam shoulders. These moderate settlements were achieved by using well-graded rockfill of good basaltic rock and heavy compaction consisting of water sluicing of the rockfill in layers (lifts) of thickness only 1.0–1.2 m with 10–12 passes by a 33-tonne vibratory roller.
- The rockfill compression modulus obtained by back-calculations based on vertical settlements measured during construction was much higher than the modulus derived from the laboratory tests. This indicates that the field compaction obtained by vibratory rollers was very effective. The roller compaction also increases the lateral confining stresses that have significant effect on the rockfill and gravel behaviour.
- Figure 10.30 shows the comparison of the measured and computed lateral (horizontal) displacements of the core in the maximum cross-section A since start of reservoir impounding (see Figure 10.25). The measured maximum lateral displacement of the core after impoundment was 86 mm in the downstream direction for a reservoir level 3 m below normal water level (NWL). The maximum lateral core displacement was at a level about 30 m above the core plinth on top of the 40-m high concrete monolith in the bottom of the canyon. At the same reservoir level, the top of the core had moved 10 mm in the upstream direction. For high reservoir levels, the 3D analysis gave results in good agreement with the measured

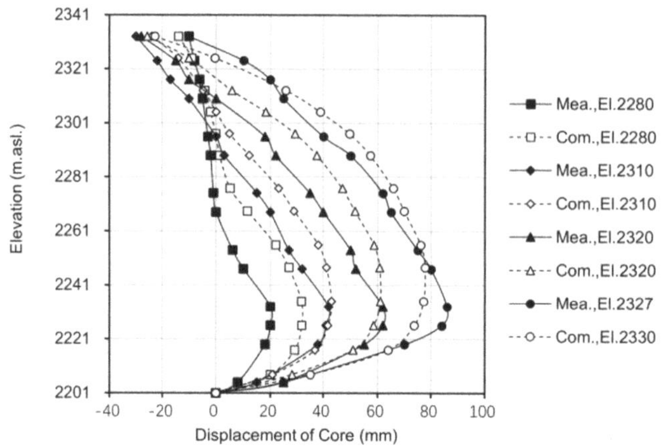

Figure 10.30 Comparison of measured and computed lateral core displacements normal to the core in cross-section A at four different reservoir levels (positive towards downstream).

values. The computed deflected shape of the core over the height agreed well with that measured. Figure 10.31 shows the comparison between 3D and 2D analyses of computed lateral displacements over the full height of the core in cross-section A since start of reservoir impounding. The 2D analysis gave maximum core displacement in the downstream direction almost twice that computed with 3D geometry and analysis.

- The comparison of computed results by 3D and 2D analyses showed that the 2D analyses gave a maximum vertical settlement at EoC nearly twice that computed by the 3D analysis. That shows the significant effect of the narrow canyon geometry. The 2D analysis computed

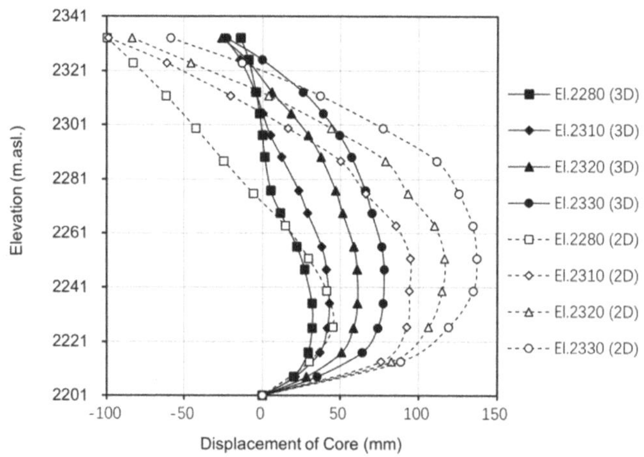

Figure 10.31 Comparison between 3D and 2D analyses of computed lateral core displacements normal to the core in cross-section A since start of reservoir impounding (positive towards downstream, 3D: solid lines; 2D: stippled lines).

maximum horizontal displacements in the upstream and downstream shoulders normal to dam axis that are about twice the displacements computed by the 3D analysis.

- The measured maximum vertical compressive strain in the asphalt core was about 4%, and the computed maximum value is about 1.7%. The computed maximum shear strain in the plane of the core was also about 4% and is located at about half of the core height up along the left abutment. These values are much smaller than the strains required to cause harmful cracking in the core as determined by laboratory tests on the asphalt concrete used in the Quxue Dam core (see Figure 3.7).
- The numerical stress-strain analyses confirm that the slender asphalt core wall acts like a relatively flexible membrane in the dam and has little effect on the overall static and dynamic dam behaviour. 3D computations in a parameter study show that changes in the asphalt concrete properties from an assumed high compression modulus to a low modulus (by using properties derived for core temperatures of 5°C, 15°C, and 25°C, respectively) have only minor effects on dam behaviour and lateral core deformations.
- The dam has been designed such that the core can accommodate the lateral displacements and deflected shape without suffering detrimental cracking or material degradation. The field monitoring and measurements taken so far confirm the satisfactory dam performance and indicate that the performance also will be satisfactory during severe earthquake loading.

Chapter 11

Comparisons with other embankment dam options

11.1 Introduction

For each dam project and construction site, there is a "competition" among various dam design options. ICOLD (2018) summarizes some of the advantages of the asphalt core embankment dam (ACED). The ACED has proved to be a competitive design option in terms of safety, construction time, and economy. The main basis for the application of impervious hydraulic asphalt concrete (HAC) in dam water barriers are the visco-elastoplastic properties. They give the barrier flexible and ductile behaviour and stress-relaxation and self-sealing properties if any cracks should occur. An added advantage is the uncomplicated and robust construction procedure and the speed of construction.

The relative merits of ACEDs compared to other types of embankment dams are briefly discussed in this chapter. Among the many important aspects to be considered during dam type selection are site topography and geology, seismicity, spillway location, weather conditions during construction, total construction time until reservoir is full/useful, consequences of dam overtopping if it should occur during dam construction, earthquake resistance, and environmental challenges caused by the construction.

11.2 ACED compared to earth core embankment dam (ECED)

When suitable, impervious earth core and filter materials are available within a reasonable transportation distance, other designs can generally not compete with the earth core in terms of economy. Furthermore, an ECED with a wide earth core and wide high-quality filters provides good dam earthquake resistance, even in cases where potential tectonic fault movements may take place in the dam foundation.

However, if the available earth core material has a high clay content, a water content significantly higher than the optimum, and the dam is to be constructed in a region with heavy precipitation during long or unpredictable periods of the year, the construction schedule becomes uncertain and satisfactory core quality control difficult to achieve. The earth core option may have to be abandoned. The Yele Dam in China represents such a case as described in Section 10.4.

If the available earth core borrow material is in a dry state in an arid region, it may be difficult to excavate the earth material and mix in water to achieve a uniform core material with optimal water content, sufficiently low permeability, and a stable soil structure with respect to internal erosion.

DOI: 10.1201/9781315177786-11

For high dams, the earth core material available may have to be modified by mixing in gravels to stiffen and strengthen the core and reduce core settlements and dam deformations. The mixing is a difficult procedure that takes time and extra costs.

The base of an earth core is much wider than the base of the slender asphalt core. Therefore, the foundation preparation and treatment under the earth core is more comprehensive than for the asphalt core founded on a concrete plinth. The core foundation area to be treated for the ACED may be approximately of 1/5 to 1/10 of the area to be treated for an ECED. On the contrary, However, the narrow core trench for the ACED may have to go deeper than the trench for the earth core to get the plinth founded on a level where the foundation quality is satisfactory.

The earth core can only be constructed in periods without rain and with positive temperatures. The asphalt core construction is much less sensitive to weather conditions, and the construction season can be significantly longer than for the ECED. Construction of the asphalt core must only stop during periods of heavy rain or snow and with temperatures below approximately –5°C to –10°C. However, the temperature control of the asphalt concrete mix must be continuous and the core surface must be dry and hot before the next layer can be placed and compacted.

In recent years, the increased concerns about internal erosion in earth cores have led to very stringent filter design and construction criteria, and more processing of filter materials is required. Natural gravel and sand may be preferable to crushed angular rock particles and a mixture of crushed rock particles and natural gravel may be the optimum design for a critical filter.

Milligan (2003) discussed uncertainties related to the engineering and construction of ECEDs. He emphasized the importance of satisfactory field placement and compaction of the earth core material and the placement of broadly graded filter materials to prevent segregation and to ensure the integrity of the dam. If the earth core is to be founded on jointed or fractured rock, it is mandatory to treat the rock surface with grout or concrete and use properly placed protective filters to avoid erosion of core material into open cracks. This aspect has too often been overlooked, even in fairly recent dams.

In addition to the required particle gradation of critical filters that depend on the properties of the earth core material to be protected, there are additional requirements concerning investigations whether the filter material can "hold a crack" if cracks should develop due to static embankment deformations or earthquake loadings. This has led to further specifications for acceptable filters to be used in ECEDs. If cracks should develop and stay open, they significantly reduce the effects of the protective filters. This happens if the content of fines in the filter is high or if the filter material contains minerals, for instance, carbonitic materials, that promote cohesion and cementation.

In the case of possible flooding and core overtopping during construction, the earth core design is more vulnerable to damage than a dam with an asphalt core, and it takes longer to start the construction up again after an accidental, temporary overtopping.

An embankment dam with a wide earth core requires more gentle exterior dam slopes and a larger embankment volume than the dam with an asphalt core.

Scarps left in the landscape caused by earth core material borrows are of environmental concern and may, in some cases, become an important factor in the decision process among potential dam options.

For the construction of the asphalt core itself, the main dam contractor may engage a specialty sub-contractor with a core placement machine (paver), or the main contractor may rent such a paver and train the staff to also do that part of the project. Both these approaches to the core construction have been used for recent ACEDs. In either case, the same technical specifications for the asphalt concrete mix design, placement, compaction, and quality control must be followed and satisfied.

11.3 ACED compared to concrete face rockfill dam (CFRD)

The concrete face dam option requires that the upstream dam shoulder should consist of well-compacted rockfill and gravel to limit the displacements, deflections, and deformations of the concrete face. Without such rockfill material available, the concrete face dam is probably not the best option.

Experience shows that leakages often occur through opening of joints and cracks in the concrete facing, especially in the vicinity of the perimeter joint along the upstream dam toe. However, in a properly designed CFRD, with gradually coarser rockfill material and increasing permeability from upstream to downstream, even large leakages do not endanger the overall dam stability if the dam has been provided with ample downstream drainage capacity and rockfill toe stability.

In several cases, a significant reduction in leakage through the concrete facing has been achieved by simply dumping silt and sand in the reservoir just upstream of the leakage locations. The silty sand migrates into the cracks and gradually reduces the leakage by clogging. This only occurs if the sand–gravel zone under the concrete facing is designed and built to function as a cohesionless filter arresting further migration of the crack-stopper material. As some leakage through opened joints and cracks in the lower part of the concrete facing has been considered to be "inevitable", placing a layer (wedge) of silt and sand on the upstream side of the lower part of the concrete facing has become a practical measure included in the basic CFRD design. In other cases, the lower part of the facing has been covered by a geomembrane.

After many years of successful application and increasing popularity of this type of embankment dam, some very high CFRDs built in narrow valleys suffered from the development of excessive compressive strains in the lower part of the facing. The compressive stresses and strains caused concrete crushing and shear crack formation during or shortly after end of construction (e.g., Pinto, 2006; Johannesson, 2007a; Johannesson, 2007b; Sabrinho et al., 2007). It has been concluded that the cracking and crushing of the facing were caused by vertical settlements and lateral displacements of the rockfill, including creep deformations, towards the bottom of the narrow valley causing added normal and shear strains in the concrete facing (e.g., Pinto, 2009). To avoid or limit such cracking, special design considerations are required for the facing by designing compressible joints between face slabs to relieve the high compressive stresses. Furthermore, the concrete facing should preferably not be placed on the upstream face until much of the anticipated post-construction creep deformations in the rockfill have taken place. It has been suggested for high CFRDs that at least ½ year of post-construction creep should be allowed before placing the concrete face slabs.

In contrast to the relatively brittle CFRD facing, the asphalt core in the ACED has no joints, and it exhibits viscoelastic and ductile behaviour. It has the ability to relieve stress concentrations and to self-heal fissures and cracks. There is no case of reported leakage through cracks or excessive shear dilation in an asphalt core, which may be considered to be maintenance-free. The only leakage that has been occurring is under the concrete plinth due to inadequate attention to grouting jointed and fractured rock at that location and was remedied by post-construction grouting.

A vertical, centrally located asphalt core can tolerate foundation settlements and embankment deformations due to static and earthquake loading better than an inclined upstream concrete facing. One can therefore accept the use of lower quality and less well-compacted alluvium or rockfill than for CFRDs, and this was used for, e.g., Finsterbach ACED (Tschernutter, 1997); Holmvatn and Storglomvatn ACEDs (Høeg et al., 2007); and Jinfeng ACED (Liao et al., 2023).

The plinth for a centrally located asphalt core is shorter than for the CFRD and is simpler in design and construction than the perimeter plinth for the CFRD as the asphalt core in most cases is vertical while the concrete facing is inclined.

An experienced contractor is required to construct a concrete facing properly, with attention paid to construction details such as water stops, steel reinforcement, and joints in the slab and along the important perimeter joint at the plinth. The whole construction process for the seamless asphalt core, including mixing, transporting, placing, and compaction is mechanized, and the core quality control procedures are well defined and easily performed. Construction workers on the steep upstream CFRD slope (often about 1V:1.3H) are exposed to a much higher occupational hazard than that experienced by the workers engaged in asphalt core construction.

Any unstable rock blocks or local slopes in the abutment above the perimeter plinth of the concrete facing must be removed or anchored to eliminate potential hazards to the workers and damages to the facing by rockfalls. A steep abutment above the ACED is of less concern as the asphalt core is located in the central region of the dam. The required bank treatment works for a CFRD is more extensive than for an ACED at the same location. In some cases, a large portion of an abutment face has to be treated for a CFRD, while only a narrow abutment strip needs to be treated for an ACED.

An advantage of the CFRD design compared to the ACED design is that the perimeter plinth construction and the grouting under the plinth may take place independently of the construction of the other zones of the embankment. The construction of the asphalt core and the adjacent transition zones can usually not start until the plinth and grouting are completed under the deepest section of the dam. If a grouting gallery is constructed near the base of the core, as discussed in Section 8.7, the foundation grouting may take place independently of the progress of embankment construction. However, the construction of a grouting gallery, if not also required for post-construction grouting, drainage, or dam monitoring, may add significant extra costs to the project. Rock excavation by blasting for a gallery may cause extra fracturing and leakage paths.

The concrete facing of the CFRD provides slope protection against wave action, while the core dam requires riprap protection Upstream slope instability caused by rapid drawdown (e.g., pump storage reservoir) may be of concern for an ACED, but usually not for a CFRD. The rockfill under the CFRD facing has the advantage of being dry (or only partly saturated) under static and earthquake loading conditions and can therefore have a steeper upstream slope than core dams. However, if the embankment is founded on a weak foundation, the external dam slopes for all options may be governed by the foundation strength and stability, and there will not be any significant differences in external slope angles among the various options.

Among existing dams, the crown portions of the core dams are observed to be significantly more robust than that for the CFRDs, which, for economy, generally relies on the use of very steep dam slopes and a concrete parapet wall that may be vulnerable, for instance during earthquake loading or sabotage potential acts of sabotage.

The ACED allows reservoir raising and storage during construction, such that the reservoir may be full by the end of embankment construction. The CFRD cannot provide storage during construction unless the facing is built in sections (and not only after the embankment is completed). Some of the more recent CFRDs have been built in sections (stages) to shorten the total construction period. This requires special attention to rockfill deformations and their effects on the facing and joints, as some areas of the facing are already in place while embankment rockfill is being raised above that level and adjacent to already constructed concrete panels (slabs). After the experiences with high CFRDs in narrow valleys (e.g., Pinto, 2009) and the serious slab cracking that occurred, one may return to the practice of constructing the facing after the embankment is completed. However, this will extend the total construction period.

As discussed in Chapter 9, the upstream facing of a CFRD is likely to be damaged due to vertical and horizontal cyclic and permanent displacements during the earthquake shaking.

The damages to the upstream facing of the Zipingpu CFRD caused by the Wenchuan earthquake in 2008 were significant, but there was no indication of dam instability. It should be noted that the reservoir water level was exceptionally low at the time of the earthquake, only about 1/3 of normal water level.

From the measured vertical and horizontal displacements and recorded dynamic response of the Zipingpu CFRD, it may be inferred that a flexible and ductile asphalt core would not have been significantly damaged if the ACRD option had been selected at that dam site. A more flexible facing with more compressible joints would have adjusted better to the rockfill deformations that took place in the Zipingpu CFRD. The geomembrane-faced embankment dam option will probably be selected more often at seismic sites in the future to take advantage of the membrane flexibility without cracking and rupture.

There is no maintenance required for an interior core as opposed to an external facing. There is no case reported of an asphalt core dam suffering from leakage due to cracking of the core. If such leakage should occur, the method of repair would be to drill through the upstream fine transition zone (Zone 2) and grout the critical location.

In an article by Materón and Freitas (2023), there is a discussion of the relative merits of the concrete face rockfill dam (CFRD) option and the ACED option compared to other types of embankment dams. An advantage of the options with upstream facing is that if the facing is damaged by flood or earthquake loading, the locations of critical leakage may be defined and repaired (short-term) relatively easily. For long-term (permanent) repair, the reservoir may temporarily have to be lowered below the level of damage.

For the ACED, the probability of damage to the water barrier is much lower than for the dam with upstream facing, as discussed in previous chapters. However, the process of pinpointing the location of the core damage and doing the required repair work is more cumbersome and difficult. Short-term repair may be achieved by drilling (vertically) through the upstream fine transition zone (Zone 2) and grout the critical location. As for the CFRD, for long-term (permanent) repair of the ACED, the reservoir may temporarily have to be lowered below the level of damage.

References

Adikari, G. S. N., Valstad, T., Kjærnsli, B., and Høeg, K. 1988. Behaviour of Storvatn Dam, Norway – a case of prediction versus performance. In *Proceedings 5th Australia-New Zealand Conference on Geomechanics*, August, pp. 86–92,Sydney, Australia.

Akhtarpour, A., and Khodaii, A. 2009. Nonlinear numerical evaluation of dynamic behaviour of an asphaltic concrete core rockfill dam (a case study), *Journal of Seismology and Earthquake Engineering (JSEE)*, 11(3):143–152.

Akhtarpour, A., and Khodaii, A. 2013. Experimental study of asphaltic concrete dynamic properties as an impervious core in embankment dams, *Construction and Building Materials,* 41(2013):319–334. https://doi.org/10.1016/j.conbuildmat.2012.11.044

Alicescu, V., Tournier, J. P., Kara, R., and Rosculet, D. 2015. Construction of La Romaine Complex in Northern Quebec: six years of great accomplishments; Behaviour of asphalt core dams. In *Canadian Dam Association Annual Conference*, October, Mississauga, Ontario.

Arnevik, A., Kjærnsli, B., and Walbö, S. 1988. The Storvatn Dam-a rockfill dam with a central core of asphaltic concrete. *The 16th ICOLD Congress,* R.9-Q.61, San Francisco, CA.

Asphalt Institute. 1979. *Mix Design Method for Asphalt Concrete and Other Hot-Mix Types*. Manual Series No. 2, The Asphalt Institute, College Park, MD.

ASTM. 2005. Standard Practice for Effect of Water on Bituminous-Coated Aggregate Using Boiling Water. D 3625-96.

Baghaee Moghaddam, T., and Haaj, H. 2018. Application of compressible packing model for optimization of asphalt concrete design, *Construction and Building Materials*, 159(2018):530–539. https://doi.org/10.1016/j.conbuildmat.2017.11.004

Baziar, M. H., Salemi, Sh., and Merrifield, C. M. 2009. Dynamic centrifuge model tests on asphalt-concrete core dams, *Géotechnique*, 59(9):763–771. https://doi.org/10.1680/geot.7.00004

Baziar, M. H. Salemi, Sh., and Heidari, T. 2006. Analysis of Earthquake Response of an Asphalt Concrete Core Embankment Dam, *International Journal of Civil Engineering*, 4(3):192–211.

Baziar, M. H., Salemi, Sh., Noorzad, A., and Ghanuad, Z. 2003. Evaluation of earthquake response of the 15th Khordad Earth Dam. In *Proceeding 21st ICOLD Congress*, Vol.3, Q.83, R.56, pp. 1011–1024, Montreal, Canada.

Belloni, L., Ulisse, C., and Valenti, D. 1988. Behaviour of an asphaltic membrane under seismic loading. In *Proceeding16th ICOLD Congress*, Vol.2, Q.61, R., pp. 1053–1073, San Francisco, CA.

Bray, J. D., and Macado, J. 2019. Procedure for estimating shear-induced seismic slope displacement for shallow crustal earthquakes, *Journal of Geotechnical and Geoenvironmental Engineering*.145(12) https://doi.org/10.1061/(ASCE)GT.1943-5606.0002143

Bray, J. D., and Travasarou, T. 2007. Simplified procedure for estimating earthquake-induced deviatoric slope displacements, *Journal of Geotechnical and Geoenvironmental Engineering*, 133(40):381–392.

Charles, J. A. 1998. Hydraulic fracture of bituminous cores of rockfill dams, Technical Report, September 1998, Building Research Establishment (BRE), Centre for Ground Engineering and Remediation.

Creegan, P. J., and Monismith, C. L. 1996. *Asphalt Concrete Water Barriers for Embankment Dams*. American Society of Civil Engineers (ASCE) Press, New York.

Cruz, P. T., Materon, B., and Freitas, M. 2009. *Concrete Face Rockfill Dams*. CRC Press, Leiden, The Netherlands.

Ding, P. 1982. Study of aggregate gradation of asphalt impervious layer. *Technology of Hydraulic Asphalt Concrete*, (1):14–27 (in Chinese).

Ding, P., Sun, Z., and Wang, W. 1991. Some problems of asphalt mix design for hydraulic structure. *Journal of Hydraulic Engineering*, (1):19–27 (in Chinese).

European Standard. 2020a. Bituminous Mixtures-Test Methods-Part 11: Determination of the Affinity Between Aggregate and Bitumen. EN 12697-11; February.

European Standard. 2020b. Bituminous Mixtures-Test Method for Hot Mix Asphalt-Part 23: Determination of Indirect Tensile Strength of Bituminous Specimens. EN 12697-23; July

Feizi-Khankandi, S., Ghalandarzadeh, A., Mirghasemi, A. A., and Høeg, K. 2009. Seismic analysis of the Garmrood embankment dam with asphaltic concrete core. *Soils and Foundations. Japanese Geotechnical Society*, 49(2):153–166. https://doi.org/10.3208/sandf.49.153

Feizi-Khankandi, S., Mirghasemi, A. A., Ghalandarzadeh, A., and Høeg, K. 2008. Cyclic triaxial tests on asphalt concrete as a water barrier for embankment dams. *Soils and Foundations*, 48(3):319–332. https://doi.org/10.3208/sandf.48.319

Fell, R., MacGregor, P., Stapledon, P., Bell, G., and Foster, M. 2015. *Geotechnical Engineering of Dams*. Taylor & Francis, London

Feng, W. 2018. Research on torsional shear test of hydraulic asphalt concrete. PhD thesis, Xi'an University of Technology, China.

Feng, S., Wang, W., Hu, W., Deng, Y., Yang, J., Wu, S., Zhang, C., and Høeg, K. 2020a. Design and performance of the Quxue asphalt-core rockfill dam, *Soils and Foundations*, 60(2020):1036–1049. https://doi.org/10.1016/j.sandf.2020.06.008

Feng, S., Wang, W., Tan, H., Xue, M., and Høeg, K. 2020b. Effects of acidic water on hydro-mechanical properties of asphalt concrete, *Case Studies in Construction Materials*, 13(2020):e00445. https://doi.org/10.1016/j.cscm.2020.e00445

Feng, S., Wang, W., Hu, K., and Høeg, K. 2020c. Stress-strain-strength behavior of asphalt core in embankment dams during construction, *Construction and Building Materials*, 259(2020):119706. https://doi.org/10.1016/j.conbuildmat.2020.119706

Gallacher, D. 1988. Asphaltic concrete core at the Megget Dam. In *Transactions, 16th International Congress on Large Dams*, R.39-Q.61, pp. 707–731, San Francisco, CA.

Gazetas, G., and Dakoulas, P. 1992. Seismic analysis and design of rockfill dams: state of the art, *Soil Dynamics and Earthquake Engineering*, 2:27–61.

Ghanooni, M. S., and Roosta, R. M. 2002. Seismic analysis and design of asphaltic concrete core embankment dams, *Hydropower and Dams*, 9(6):75–78.

Griffiths, D. V., and Lane, P. A. 1999. Slope stability by finite elements, *Geotechnique*, 49(3):387–403.

Guan, Z. (Editor). 2014. *Embankment Dams-Handbook of Hydraulic Structure Design*. China Water Power Press, Beijing, China (in Chinese).

Guo, L., He, J., and Yang, H. 2020. Study on wind-proof structure with filling height difference between dam body and core wall under crosswind, *Advances in Science and Technology of Water Resources*, 40(5):89–94 (in Chinese).

Haas, H. 1983. *Zur eignung und optimierung von asphaltbeton für kerndichtungen in staudämmen*, 3, Bitumen, Heft, pp. 97–106.

Han, B., Zdravkovic, L., Kontoe, S., and Taborda, D. M. G. 2016. Numerical investigation of the response of the Yele rockfill dam during the 2008 Wenchuan earthquake, *Soil Dynamics and Earthquake Engineering*, 88(2016):124–142.

Han, X., Hu, Z., Yu, L., Pang, Y., She, H., Zhang, L., Wang, X., and Qi, C. 2023. Dynamic characteristics of asphalt concrete as an impervious core in embankment dams under various temperatures and stress states, *Materials*, 16(2023):6529. https://doi.org/10.3390/ma16196529

He, Y., Xiong, K., and Li, P. 2008. Seismic response features of Yele dam in Wenchuan earthquake. In *14th World Conference on Earthquake Engineering*, October 12–17, Beijing, China.

Hinks, L. J. (Editor). 2023. *Earthquake Engineering for Dams and Reservoirs*. Emerald Publishing Limited, Leeds.

Hoek, E., Kaiser P. K., and Bawden, W. E. 1995. *Support of Underground Excavations in Hard Rock*, Balkema, Rotterdam.

Høeg, K. 1993. *Asphaltic Concrete Cores for Embankment Dams*. Stikka Press, Oslo.

Høeg, K. 1995. Transverse cracking in embankment dams-a literature review and finite element study, Report 532060-1, Norwegian Geotechnical Institute (NGI); March.

Høeg, K., Valstad, T., Kjærnsli, B., and Ruud, A. M. 2007. Asphalt core embankment dams: recent case studies and research, *Hydropower and Dams*, 13(5):112–119.

Hu, C. L., Huang, L. S., and Li, Y. H. 2001. Study on the deformation properties of asphaltic concrete core wall in an embankment dam, *Journal of Wuhan University of Technology*, 23(9):47–49.

ICOLD. 1982. *Bituminous Cores for Earth and Rockfill Dams*. Bulletin 42. ICOLD, Paris, France.

ICOLD. 1992. *Bituminous Cores for Fill Dams-State of the Art*. Bulletin 84. ICOLD, Paris, France.

ICOLD. 1999. *Embankment Dams with Bituminous Concrete Facing*. Bulletin 114. ICOLD, Paris, France.

ICOLD. 2018. *Asphalt Concrete Cores for Embankment Dams*. Bulletin 179. ICOLD, Paris, France.

ICOLD. 2005. *Dam Foundations. Geologic Considerations. Investigation Methods. Treatment. Monitoring*. Bulletin 129, ICOLD, Paris, France.

Jafarzadeh, F., and Javaheri, H. 2003. Dynamic analysis of rockfill dams considering elasto-plastic behaviour and three-dimensional canyon effects. In *Proceedings of the 21st ICOLD Congress*, Vol.3, Q.83, R.9, pp. 109–129, Montreal, Canada.

Jiao, Y., Ren, G., Peng, W., Gu, X., and Wang, N. 2020. Aseismic measures for asphalt core dams by dynamic centrifuge modelling tests, *Chinese Journal of Geotechnical Engineering*, 42(Supp.1):167–171 (in Chinese).

Johannesson, P. 2007a. Assessing the performance of high CFRDs: rockfill properties and creep, *Hydropower and Dams*, 14(5):120–127.

Johannesson, P. 2007b. Slab performance on a few CFRDs and suggested improvements in slab design. In *Proceedings of the International Symposium on Key Techniques for Dam Construction*, Yichang, China, May; 22–24.

Jones, G. A., White, A. C., and Schreuder, W. J. 1999. Design and construction of the asphaltic concrete core at Ceres Dam. In *Geotechnics for Developing Africa* (Blight, G. E., Fourie, A. B. and Wardle, G. R. (eds)). CRC Press, London, SRK-Report, June.

Karlsen, L. E., Krogh, R. M., and Johansen, P. M. 1995. Dam construction for Norway's largest hydropower reservoir, *Hydropower and Dams*, 2(3):27–30

Kawamura, H. 1989. The follow-up investigation on the durability of the asphalt concrete facing for the Futaba Dam, *Water Soils*, 78(1989):41–53 (in Japanese).

Kim, Y. R. (Editor). 2009. *Modeling of Asphalt Concrete*. ASCE Press, Reston, VA.

Kjærnsli, B., Moum, J., and Torblaa, I. 1966. Laboratory tests on asphaltic concrete for an impervious membrane on the Venemo rockfill dam. *Norwegian Geotechnical Institute Publication*, 69:17–26.

Le Coroller, A., Bienaimé, C., Herment, R., Poupart, M., Huynh, P., Crepel, J. M., Modaressi, A., and Bard, E. 1988. Approach to bituminous concrete core behaviour by finite element model computation. In *Proceedings of the 16th ICOLD Congress*, Vol.2, Q.61, R.54, pp. 997–1026, Paris, France.

Lengfeld, M., and Müller, D. 2018. Partial replacement of the Obernau dam's asphalt concrete facing, *Wasserwirteschaft*, 108(6):68–71.

Li, J., Liu, Y., Fang, C., Huang, H., Meng, F., Zhou, F., He, J., and Yang, Y. 2021. *Key Technologies and Practices of Ultra-Deep Cut-Off Walls for Deep Overburden Dam Foundations*. China Water Power Press, Beijing, China (in Chinese).

Li, N., Li, G., and Shen, Z. 2003. Dynamic properties of embankment material and behaviour of high concrete faced gravel-rockfill dam during earthquake. In *Proceedings of the 21st ICOLD Congress*, Vol.3, Q.83, R.70, pp. 1211–1227, Montreal, Canada.

Liao, D., Xu, G., Hu, G., and Peng, M. 2023. Design and performance of Jingfeng asphalt core embankment dam built with soft sandstone debris. In *Concrete Face Rockfill Dams 2023 Annual Conference*, Zhengzhou, China (in Chinese).

Liu, H., Chen, Y., Yu, T., and Yang, G. 2015. Seismic analysis of the Zipingpu concrete-faced rockfill dam response to the 2008 Wenchuan, China, Earthquake, *Journal of Performance of Constructed Facilities*, 29(5). https://ascelibrary.org/doi/10.1061/(ASCE)CF.1943-5509.0000506

Liu, Y., Ye, Y., Luo, W., and Peng, H. 2022. Modeling the behavior of an aggregate skeleton during static creep of an asphalt mixture based on a three-dimensional mesoscale random model, *Journal of Materials in Civil Engineering*, 34(12):04022319. https://doi.org/10.1061/(ASCE)MT.1943-5533.0004470

Longtin, H., Verret, D., James, M. Grenier, S., Noel, J-F., and Hammamji, Y. 2013. Romaine-2: numerical stress-deformation analyses of the dam and dykes. In *CDA 2013 Annual Conference*, Quebec, Canada.

Materón, B., and Freitas, M. S. 2023. CFRDs and ACRDs: advancements in rockfill dams, *International Water Power and Dam Construction*, November, pp. 36–39.

Mejia, L. H. 2021. Earthquake cracking of embankment dams, 23rd Ralph B. Peck lecture, *Journal of Geotechnical and Geoenvironmental Engineering*, 150(11). https://doi.org/10.1061/JGGEFK.GTENG-12751

Meng, X, Liu, Y., Ning, Z., Dong, J., Liang, G., and Wang, Y. 2023. Fatigue performance of hydraulic asphalt concrete under uniaxial constant-amplitude tensile cycle loading, *Journal of Building Engineering*, 80(2023):108070. https://doi.org/10.1016/j.jobe.2023.108070

Menq, F. Y., and Stokoe, K. H. 2003. Linear dynamic properties of sand gravelly soils from large-scale resonant tests. In *Deformation Characteristics of Geomaterials. Proceedings of the 3rd International Symposium on Deformation Characteristics of Geomaterials*, Lyon, France (Di Benedetto, H., Doanh, T., Geoffroy, H. and Sauzéat, C. (eds)). Balkema, Lisse, the Netherlands, pp. 63–71.

Milligan, V. 2003. Some uncertainties in embankment dam engineering. 38th Terzaghi Lecture, *ASCE Journal of Geotechnical and Geoenvironmental Engineering*, 129(9):785–797.

Moiseev, I. S., Yakolev, N. N., Goldin, A. L., Gorelink, L. V., and Radchenko, V. G. 1988. Rockfill dams with asphalt diaphragms. In *16th ICOLD Congress*, R.62, Q.61, San Francisco, CA.

Nakamura, Y., Okumura, T., Narita, K., and Ohne, Y. 2004. Improvement of impervious asphalt mixture for high ductility against earthquake excitation. In *Proceedings of the 4th International Conference on Dam Engineering*, Nanjing, China, 18–20 October 2004, pp. 647–656. A.A. Balkema, Netherlands.

NEA. 2019. Test Code for Hydraulic Asphalt Concrete DL/T 5362-2018, National Energy Administration of China, China Electric Power Press, Beijing, China (in Chinese).

NEA. 2023. Code for Design of Asphalt Concrete Facings and Cores for Embankment Dams NB/T 11015-2022, National Energy Administration of China, China Water & Power Press, Beijing, China (in Chinese).

Newmark, N. M. 1965. Effects of earthquakes on dams and embankments, *Géotechnique*, 15(2):139–160.

Ning, Z., Liu, Y., Wang, W., Dong, J., Meng, X., and Wang, Q. 2024. Crack propagation and non-uniform deformation analysis in hydraulic asphalt concrete: insights from X-ray CT scanning and digital volume correlation, *Construction and Building Materials*, 411(2024):134704. https://doi.org/10.1016/j.conbuildmat.2023.134704

Pinto, N. de S. 2006. Cracking of concrete slabs for the Barra Grande and Campos Novos dams in Brazil. In *Discussion Presented at the 22nd ICOLD Congress*, Barcelona, Spain.

Pinto, N. de S. 2009. The design and construction of extra high CFRDs, *The International Journal on Hydropower and Dams*, 16(3):41–44.

Pircher, W., and Schwab, P. 1988. Design, construction and behaviour of the asphaltic concrete core wall of the Finstertal Dam. In *The 16th ICOLD Congress*, R49-Q61, pp. 901–924. San Francisco, CA.

Qin, X., Huang, X., Peng, L., and Wang, Q. 2011. Design and construction of asphalt core. Series of books for Xinjiang Xiabandi key water control project. China Water Power Press, Beijing, China (in Chinese).

Qiu, T., Wang, W., Høeg, K., Feng, S., and Zhao, R. 2021. 3D analysis of the 174-m high Quxue asphalt-core rockfill dam in a narrow canyon, *Soils and Foundations*, 61(2021):1645–1659. https://doi.org/10.1016/j.sandf.2021.10.004

Qu, M., Wang, W., and Cai, X. 2004. Experimental research on asphalt concrete impervious core wall for Yele hydropower station, *Journal of Hydroelectric Engineering*, 23(6):80–82 (in Chinese).

Rienossl, K. 1973. Embankment dams with asphaltic concrete cores - experience and recent test results. In *Transactions, 11th International Congress on Large Dams*, R.45-Q.42; pp. 801–816, Madrid, Spain.

Sabrinho, J. A., Xavier, V., Albertino, S. C., Correa, C., and Pereira, R. F. 2007. Performance and concrete repair at Campos Novos, *Hydropower and Dams*, 14(2):39–42.

Saxegaard, H. 2002. Asphalt core dams: increased productivity to improve speed of construction. *Hydropower & Dams*, 9(6):72–74.

Saxegaard, H. 2003. Crack self-healing properties of asphalt concrete: laboratory simulation. *Hydropower & Dams*, (3):106–109.

Schönian, E. 1999. *The Shell Bitumen Hydraulic Engineering Handbook*. Thomas Telford, London.

Schober, W. 1973. Considerations and investigations for the design of a rockfill dam with a 92 m high bituminous mix core. In *Transactions, 11th International Congress on Large Dams*, Q.42, R.34, pp. 575–600, Madrid, Spain.

Schwab, H., and Pircher, W. 1985. Structural behaviour of a high rockfill dam comprehensive interpretation of measurements, and conclusions on stress-strain relationships. In *Proceedings of the 15th International Congress on Large Dams (ICOLD)*, Q.56, R.67, pp. 1273-1297, Lausanne, Switzerland.

Seed, H. B. 1979. Considerations in the earthquake-resistant design of earth and rockfill dams, *Gétechnique*, 29(3):215–263.

Sembenelli, P., and Afsari-rad, V. 2015. High asphalt core for Zarema Dam. In *25th ICOLD Congress*, Q.98, R.34, pp. 526–538, Stavanger, Norway.

Smith, M. 2015. Rockfill settlement measurement and modelling of the Romaine-2 Dam during construction. In *25th ICOLD Congress*, Q.98, R.31, Stavanger, Norway.

Shoichi, ABE, Takehiko, OKUDERA, Hiroyuki, SETO. 2012. Rational design and construction of embankment and asphalt facing on the upper reservoir at Kyogoku project. In *24th ICOLD Congress*, C.2, pp. 22–43, Kyoto

Stokoe, K. H., Cox, B. R., Clayton, P., and Menq, F. 2017. NHERI TEXAS, Experimental Facility: Large-Scale Mobile shakers for natural-hazards field studies. In *Proceedings of the 16th World Conference on Earthquake Engineering*, Santiago, Chile.

Swaisgood, J. 2014. Behavior of embankment dams during earthquake, *Journal of Dam Safety*, 12(2):35–44.

Tirunas, D., and Aspen, B. 2023. Moglicé Hydropower Project, Albania – Design and Construction Experience. In *Proceedings HYDROPOWER 2018*, Gdansk, Poland, Paper No. 07.02.

Tong, J., Shen, K., Ma, T., and Zhang, J. 2022. Characterizing the tension-compression asymmetric viscoelasticity of asphalt mixture using the uniaxial tension and compression test, *Construction and Building Materials*, 350(2022):128854. https://doi.org/10.1016/j.conbuildmat.2022.128854

Tschernutter, P. 1997. Influence of soft rockfill material on a central bituminous concrete core membrane. In *Proceedings, Non-Soil Water Barriers for Embankment Dams, 17th USCOLD Lecture Series*, San Diego, CA; April.

USSD. 2022. Analysis of Seismic Deformations of Embankment Dams. Report prepared by USSD Committee on Earthquakes, Sub-Committee on Seismic Deformation Analysis of Embankment Dams, Chairman L.H. Mejia.

Valstad, T., Selness, P.B., Nadim, F., and Aspen, B. 1991. Seismic response of a rockfill dam with an asphaltic concrete core, *Journal of Water Power and Dam Construction*, 43:1–6.

Wang, W. 2004. Research on asphalt concrete core of Aikou dam. Xi'an University of Technology, Report, May (in Chinese).

Wang, W. 2008. Research on the suitability of asphalt concrete as water barrier in dams and dikes. PhD thesis, Department of Geosciences, University of Oslo, Norway.

Wang, W. 2024. A material stress-strain-time-temperature creep model for the analysis of asphalt cores in embankment dams, *Applied Sciences*, 14:3399. https://doi.org/10.3390/app14083399

Wang, W., Feng, S., and Zhang, Y. 2017a. Investigation of asphalt core-plinth connection in embankment dams, *Case Studies in Construction Materials*, 7(2017):305–316. https://doi.org/10.1016/j.cscm.2017.09.002

Wang, W., Feng, S., and Zhang, Y. 2018. Investigation of interface between asphalt core and gravel transition zone in embankment dams, *Construction and Building Materials*, 185(2018):148–155. https://doi.org/10.1016/j.conbuildmat.2018.07.078

Wang, W., and Høeg, K. 2002. Effects of compaction method on the properties of asphalt concrete for hydraulic structures, *Hydropower and Dams*, 9(6):63–71.

Wang, W., and Høeg, K. 2009. Method of compaction has significant effects on stress-strain behavior of hydraulic asphalt concrete, *Journal of Testing Evaluation*, 37(3):264–274. https://doi.org/10.1520/JTE102114

Wang, W., and Høeg, K. 2010. Developments in the design and construction of asphalt core dams, *Hydropower and Dams*, 17(3):83–91.

Wang, W., and Høeg, K. 2011. Cyclic behavior of asphalt concrete used as impervious core in embankment dams, *Journal of Geotechnical and Geoenvironmental Engineering, ASCE*, 137(5):536–544. https://doi.org/10.1061/(ASCE)GT.1943-5606.0000449

Wang, W., and Høeg, K. 2016. Simplified material model for analysis of asphalt core in embankment dams, *Construction and Building Materials*, 124(2016):199–207. https://doi.org/10.1016/j.conbuildmat.2016.07.077

Wang, W., Høeg, K., and Zhang, Y. 2010a. Design and performance of the Yele asphalt-core rockfill dam, *Canadian Geotechnical Journal*, 47(12):1365–1381. https://doi.org/10.1139/T10-028

Wang, W., Hu, K., Feng, S., and Zhao, R. 2020a. A case study on asphalt core construction rate for the Zhaobishan embankment dam, *Case Studies in Construction Materials*, 13(2020):e00418. https://doi.org/10.1016/j.cscm.2020.e00418

Wang, W., Hu, K., Feng, S., Li, G., and Høeg, K. 2020b. Shear behavior of hydraulic asphalt concrete at different temperatures and strain rates, *Construction and Building Materials*, 230(2020):117022. https://doi.org/10.1016/j.conbuildmat.2019.117022

Wang, W., Zhang, Y., Høeg, K., and Zhu, Y. 2010b. Investigation of the use of strip-prone aggregates in hydraulic asphalt concrete, *Construction and Building Materials*, 24(2010):2157–2163. https://doi.org/10.1016/j.conbuildmat.2010.04.043

Wang, W., Zhang, Y., and Xu, T. 2017b. Design and construction of the Quxue asphalt core rockfill dam, *Hydropower and Dams*, 24(4):87–92.

Wang, W., Zhang, Y., Zhu, Y., and Lan, X. 2013. Investigation on aging of asphalt concrete reservoir facing with time, *Advances in Civil Engineering Materials*, 2(1):347–359.

Wieland, M. 2003. Seismic aspects of dams, general report, Q.83 seismic aspects of dams. In *Proceedings, 21st International Congress on Large Dams*, ICOLD, Montreal, Canada.

Yasuda, N, Yamaguchi, Y., and Matsumoto, N. 2003. Dynamic deformation and strength properties of rockfill materials. In *Proceedings of the 21st ICOLD Congress*, Vol.3, Q.83, R.34, pp. 577–594, Montreal, Canada.

Zhang, Y., Høeg, K., Wang, W., and Zhu, Y. 2013.Watertightness, cracking resistance, and self-healing of asphalt concrete used as a water barrier in dams, *Canadian Geotechnical Journal*, 50(3):275–287. https://doi.org/10.1139/cgj-2011-0443

Zhang, Y., Wang, W., Du, X., and He, Z. 2008. Research on effects of limestone filler content on the behavior of hydraulic asphalt concrete, *Journal of Hydroelectric Engineering*, 5(27)(2008):73–77 (in Chinese).

Zhang, Y., Zhu, Y., Wang, W., Ning, Z., Feng, S., and Høeg, K. 2021.Compressive and tensile stress-strain-strength behavior of asphalt concrete at different temperatures and strain rates, *Construction and Building Materials*, 311(2021):125362. https://doi.org/10.1016/j.conbuildmat.2021.125362

Zhao, Y., Tang, J., and Liu, H. 2012. Construction of triaxial dynamic modulus master curve for asphalt mixtures, *Construction and Building Materials*, 37(2012):21–26. https://doi.org/10.1016/j.conbuildmat.2012.06.067

Index

Note: **Bold** page numbers refer to tables and *italic* page numbers refer to figures.